Survival of the Sickest

SURVIVAL

OF THE

SICKEST

A Medical Maverick Discovers
Why We Need Disease

DR. SHARON MOALEM

with Jonathan Prince

HarperCollins*Publishers*

This book contains advice and information relating to health care.
It is not intended to replace medical advice and should be used to supplement
rather than replace regular care by your doctor. It is recommended that
you seek your physician's advice before embarking on any medical program
or treatment. All efforts have been made to assure the accuracy of
the information contained in this book as of the date of publication.
The publisher and the author disclaim liability for any medical outcomes
that may occur as a result of applying the methods suggested in this book.

Excerpt from *The Cure at Troy: A Version of Sophocles' Philoctetes*
by Seamus Heaney. Copyright © 1990 by Seamus Heaney. Reprinted by
permission of Farrar, Straus and Giroux, LLC, and Faber and Faber Ltd.

HarperCollins*Publishers*
77–85 Fulham Palace Road,
Hammersmith, London W6 8JB

www.harpercollins.co.uk

Published by HarperCollins*Publishers* 2007
1

ISBN-10 0 00 723610 7
ISBN-13 978 0 00 723610 7

Printed and bound in Great Britain by
Clays Ltd, St Ives plc

To my grandparents
Tibi and Josephina Elizabeth Weiss,
whose lives served to teach me
the complexities of survival

CONTENTS

INTRODUCTION

This is a book about mysteries and miracles. About medicine and myth. About cold iron, red blood, and neverending ice. It's a book about survival and creation. It's a book that wonders why, and a book that asks why not. It's a book in love with order and a book that craves a little chaos.

Most of all, it's a book about life—yours, ours, and that of every little living thing under the sun. About how we all got here, where we're all going, and what we can do about it.

Welcome to our magical medical mystery tour.

WHEN I WAS fifteen years old, my grandfather was diagnosed with Alzheimer's disease. He was seventy-one. Alzheimer's—as too many people know—is a terrible disease to watch. And when

you're fifteen, watching a strong, loving man drift away almost before your eyes, it's hard to accept. You want answers. You want to
know why.

Now, there was one thing about my grandfather that always
struck me as kind of strange—he loved to give blood. And I mean
he *loved* it. He loved the way it made him feel; he loved the way it
energized him. Most people donate blood purely because it makes
them feel good emotionally to do something altruistic—not my
grandfather; it made him feel good both emotionally *and* physi-
cally. He said no matter where his body hurt, all he needed was a
good bleeding to make the aches and pains go away. I couldn't understand how giving away a pint of the stuff our lives depend on
could make someone feel so good. I asked my high school biology
teachers. I asked the family doctor. Nobody could explain it. So
I felt it was up to me to figure it out.

I convinced my father to take me to a medical library, where I
spent countless hours searching for an answer. I don't know how I
possibly found it among the thousands and thousands of books in
the library, but something steered me there. In a hunch, I decided
to plow through all the books about iron—I knew enough to know
that iron was one of the big things my grandfather was giving up
every time he donated blood. And then—bam! There it was—a
relatively unheard of hereditary condition called hemochromatosis. Basically, hemochromatosis is a disorder that causes iron to
build up in the body. Eventually, the iron can build up to dangerous
levels, where it damages organs like the pancreas and the liver;
that's why it's also called "iron overload." Sometimes, some of that
excess iron is deposited in the skin, giving you a George Hamilton
perma-tan all year long. And as we'll explore, giving blood is the
best way to reduce the iron levels in your body—all my grand-

father's blood donations were actually treating his hemochromatosis!

Well, when my grandfather was diagnosed with Alzheimer's, I had a gut instinct that the two diseases had to be connected. After all, if hemochromatosis caused dangerous iron buildups that damaged other organs, why couldn't it contribute to damage in the brain? Of course, nobody took me very seriously—I was fifteen.

When I went to college a few years later, there was no question that I was going to study biology. And there was no question that I was going to keep on searching for the link between Alzheimer's and hemochromatosis. Soon after I graduated, I learned that the gene for hemochromatosis had been pinpointed; I knew that this was the right time to pursue my hunch seriously. I delayed medical school to enter a Ph.D. program focused on neurogenetics. After just two years of collaborative work with researchers and physicians from many different laboratories we had our answer. It was a complex genetic association, but sure enough there was indeed a link between hemochromatosis and certain types of Alzheimer's disease.

It was a bittersweet victory, though. I had proved my high school hunch (and even earned a Ph.D. for it), but it did nothing for my grandfather. He had died twelve years earlier, at seventy-six, after five long years battling Alzheimer's. Of course, I also knew that this discovery could help many others—and that's why I wanted to be a physician and a scientist in the first place.

And actually, as we'll discuss more in the next chapter, unlike many scientific discoveries, this one came with the potential for an immediate payoff. Hemochromatosis is one of the most common genetic disorders in people descended from Western Europeans: more than 30 percent carry these genes. And if you know you have hemochromatosis, there are some very straightforward steps you can take to reduce the iron levels in your blood and prevent the iron buildups that can damage your organs, including the one my grandfather discovered on his own—bleeding. And as for knowing whether or not you have hemochromatosis—well, there are a couple of very simple blood tests used to make the diagnosis. That's about it. And if the results come back positive, then you start to give blood regularly and modify your diet. But you can live with it.

I do.

I WAS AROUND eighteen when I first started feeling "achy." And then it dawned on me—maybe I have iron overload like my grandfather. And sure enough, the tests came back positive. As you can imagine, that got me thinking—what did this mean for me? Why did I get it? And the biggest question of all—why would so many people inherit a gene for something potentially so harmful? Why would evolution—which is supposed to weed out harmful traits and promote helpful ones—allow this gene to persist?

That's what this book is about.

The more I plunged into research, the more questions I wanted answered. This book is the product of all the questions I asked, the research they led to, and some of the connections uncovered along the way. I hope it gives you a window into the beautiful, varied, and interconnected nature of life on this wonderful world we inhabit.

Instead of just asking what's wrong and what can be done about it, I want people to look behind the evolutionary curtain, to ask why this condition or that particular infection occurs in the first place. I think the answers will surprise you, enlighten you, and—in the long run—give all of us a chance to live longer, healthier lives.

We're going to start by looking at some hereditary disorders. Hereditary disorders are very interesting to people like me who study both evolution and medicine—because common conditions that are only caused by inheritance should die out along the evolutionary line under most circumstances.

Evolution likes genetic traits that help us survive and reproduce—it doesn't like traits that weaken us or threaten our health (especially when they threaten it before we can reproduce). That preference for genes that give us a survival or reproductive advantage is called natural selection. Here are the basics: If a gene produces a trait that makes an organism less likely to survive and reproduce, that gene (and thus, that trait) won't get passed on, at least not for very long, because the individuals who carry it are less likely to survive. On the other hand, when a gene produces a trait that makes an organism better suited for the environment and more likely to reproduce, that gene (and again, that trait) is more likely to get passed on to its offspring. The more advantageous a trait is, the faster the gene that produces it will spread through the gene pool.

So hereditary disorders don't make much evolutionary sense at first glance. Why would genes that make people sick still be in the gene pool after millions of years? You'll soon find out.

From there, we're going to examine how the environment of our ancestors helped to shape our genes.

We're also going to look at plants and animals and see what we can learn from their evolution—and what effect their evolution has had on ours. We're going to do the same thing with all the other living things that inhabit our world—bugs, bacteria, fungi protozoa, even the quasi-living, that vast collection of parasitic viruses and genes we call transposons and retrotransposons.

By the time we're through, you'll have a new appreciation for the amazing collection of life on this amazing planet of ours. And—I hope—a new sense that the more we know about where we came from, whom we live with, and where *they* came from, the more we can do to control where we want to go.

BEFORE YOU DIVE in, you need to discard a few preconceptions that you may have picked up before you picked up this book.

First of all, you are not alone. Right now, whether you're lying in bed or sitting on the beach, you're in the company of thousands of living organisms—bacteria, insects, fungi, and who knows what else. Some of them are inside you—your digestive system is filled with millions of bacteria that provide crucial assistance in digesting food. Constant company is pretty much the status quo for every form of life outside a laboratory. And a lot of that life is interacting as organisms affect one another—sometimes helpfully, sometimes harmfully, sometimes both.

Which leads to the second point—evolution doesn't occur on its own. The world is filled with a stunning collection of life. And every single living thing—from the simplest (like the schoolbook favorite, the amoeba) to arguably the most complex (that would be us)—is hardwired with the same two command lines: survive and reproduce. Evolution occurs as organisms try to improve the odds for survival and reproduction. And because, sometimes, one organism's survival is another organism's death sentence, evolution in any one species can create pressure for evolution in hundreds or thousands of other species. And that, when it happens, will create evolutionary pressure in hundreds or thousands of other species.

That's not even the whole story. Organisms' interaction with

one another isn't the only influence on their evolution; their interaction with the planet is just as important. A plant that thrives in a
tropical swamp has got to change or die when the glaciers slide
into town. So, to the list of things that influence evolution, add all
the changes in earth's environment, some massive, some minor,
that have occurred over the 3.5 billion years (give or take a few
hundred million) since life first appeared on the planet we call
home.

So to be crystal clear: everything out there is influencing the
evolution of everything else. The bacteria and viruses and parasites
that cause disease in us have affected our evolution as we have
adapted in ways to cope with their effects. In response they have
evolved in turn, and keep on doing so. All kinds of environmental
factors have affected our evolution, from shifting weather patterns
to changing food supplies—even dietary preferences that are
largely cultural. It's as if the whole world is engaged in an intricate,
multilevel dance, where we're all partners, sometimes leading,
sometimes following, but always affecting one another's movements—a global, evolutionary Macarena.

Third, mutation isn't *bad*; more to the point, it's not only good
for X-Men. Mutation just means change—when mutations are
bad, they don't survive; when they're good, they lead to the evolution of a new trait. The system that filters one from the other is
natural selection. When a gene mutates in a way that helps an organism survive and reproduce, that gene spreads through the gene
pool. When it hurts an organism's chance of survival or reproduction, it dies out. (Of course, good is a matter of perspective—a mutation that helps bacteria develop antibiotic resistance isn't good
for *us*, but it *is* good from the bacteria's point of view.)

Finally, DNA isn't destiny—it's history. Your genetic code

doesn't determine your life. Sure, it shapes it—but exactly how it shapes it will be dramatically different depending on your parents, your environment, and your choices. Your genes are the evolution- ary legacy of every organism that came before you, beginning with your parents and winding all the way back to the very beginning. Somewhere in your genetic code is the tale of every plague, every predator, every parasite, and every planetary upheaval your ances- tors managed to survive. And every mutation, every change, that helped them better adapt to their circumstances is written there.

The great Irish poet Seamus Heaney wrote that once in a life- time hope and history can rhyme. Evolution is what happens when history and change are in rhyme.

> *if there's fire on the mountain*
> *or lightning and storm*
> *and a god speaks from the sky.*
> *That means someone is hearing*
> *the outcry and the birth-cry*
> *of new life at its term.*

Survival of the Sickest

CHAPTER 1

IRONING IT OUT

A ran Gordon is a born competitor. He's a top financial executive, a competitive swimmer since he was six years old, and a natural long-distance runner. A little more than a dozen years after he ran his first marathon in 1984 he set his sights on the Mount Everest of marathons—the Marathon des Sables, a 150-mile race across the Sahara Desert, all brutal heat and endless sand that test endurance runners like nothing else.

As he began to train he experienced something he'd never really had to deal with before—physical difficulty. He was tired all the time. His joints hurt. His heart seemed to skip a funny beat. He told his running partner he wasn't sure he could go on with training, with running at all. And he went to the doctor.

Actually, he went to *doctors*. Doctor after doctor—they couldn't account for his symptoms, or they drew the wrong conclusion. When his illness left him depressed, they told him it was stress and recommended he talk to a therapist. When blood tests revealed a liver problem, they told him he was drinking too much. Finally, after three years, his doctors uncovered the real problem. New tests revealed massive amounts of iron in his blood and liver—off-the-charts amounts of iron.

Aran Gordon was rusting to death.

HEMOCHROMATOSIS IS A hereditary disease that disrupts the way the body metabolizes iron. Normally, when your body detects that it has sufficient iron in the blood, it reduces the amount of iron absorbed by your intestines from the food you eat. So even if you stuffed yourself with iron supplements you wouldn't load up with excess iron. Once your body is satisfied with the amount of iron it has, the excess will pass through you instead of being absorbed. But in a person who has hemochromatosis, the body always thinks that it doesn't have enough iron and continues to absorb iron unabated. This iron loading has deadly consequences over time. The excess iron is deposited throughout the body, ultimately damaging the joints, the major organs, and overall body chemistry. Unchecked, hemochromatosis can lead to liver failure, heart failure, diabetes, arthritis, infertility, psychiatric disorders, and even cancer. Unchecked, hemochromatosis will lead to death.

For more than 125 years after Armand Trousseau first described it in 1865, hemochromatosis was thought to be extremely rare. Then, in 1996, the primary gene that causes the condition was isolated for the first time. Since then, we've discovered that the gene for hemochromatosis is the most common genetic variant in people of Western European descent. If your ancestors are Western European, the odds are about one in three, or one in four, that you carry at least one copy of the hemochromatosis gene. Yet only one in two hundred people of Western European ancestry actually have hemochromatosis disease with all of its assorted symptoms. In genetics parlance, the degree that a given gene manifests itself in an individual is called penetrance. If a single gene means everyone who carries it will have dimples, that gene has very high or complete penetrance. On the other hand, a gene that requires a host of other circumstances to really manifest, like the gene for hemochromatosis, is considered to have low penetrance.

Aran Gordon had hemochromatosis. His body had been accumulating iron for more than thirty years. If it were untreated, doctors told him, it would kill him in another five. Fortunately for Aran, one of the oldest medical therapies known to man would soon enter his life and help him manage his iron-loading problem. But to get there, we have to go back.

WHY WOULD A disease so deadly be bred into our genetic code? You see, hemochromatosis isn't an infectious disease like malaria, related to bad habits like lung cancer caused by smoking, or a viral invader like smallpox. Hemochromatosis is inherited—and the gene for it is very common in certain populations. In evolutionary terms, that means we asked for it.

Remember how natural selection works. If a given genetic trait makes you stronger—especially if it makes you stronger before you have children—then you're more likely to survive, reproduce, and pass that trait on. If a given trait makes you weaker, you're less likely to survive, reproduce, and pass that trait on. Over time, species "select" those traits that make them stronger and eliminate those traits that make them weaker.

So why is a natural-born killer like hemochromatosis swimming in our gene pool? To answer that, we have to examine the relationship between life—not just human life, but pretty much all life—and iron. But before we do, think about this—why would you take a drug that is guaranteed to kill you in forty years? One reason, right? It's the only thing that will stop you from dying tomorrow.

JUST ABOUT EVERY form of life has a thing for iron. Humans need iron for nearly every function of our metabolism. Iron carries oxygen from our lungs through the bloodstream and releases it in the body where it's needed. Iron is built into the enzymes that do most of the chemical heavy lifting in our bodies, where it helps us to detoxify poisons and to convert sugars into energy. Iron-poor diets and other iron deficiencies are the most common cause of anemia, a lack of red blood cells that can cause fatigue, shortness of breath, and even heart failure. (As many as 20 percent of menstruating women may have iron-related anemia because their monthly blood loss produces an iron deficiency. That may be the case in as much as half of all pregnant women as well—they're not menstruating, but the passenger they're carrying is hungry for iron too!) Without enough iron our immune system functions poorly, the skin gets pale, and people can feel confused, dizzy, cold, and extremely fatigued.

Iron even explains why some areas of the world's ocean are crystal clear blue and almost devoid of life, while others are bright green and teeming with it. It turns out that oceans can be seeded with iron when dust from land is blown across them. Oceans, like parts of the Pacific, that aren't in the path of these iron-bearing winds develop smaller communities of phytoplankton, the single-celled creatures at the bottom of the ocean's food chain. No phyto-plankton, no zooplankton. No zooplankton, no anchovies. No anchovies, no tuna. But an ocean area like the North Atlantic, straight in the path of iron-rich dust from the Sahara Desert, is a green-hued aquatic metropolis. (This has even given rise to an idea to fight global warming that its originator calls the Geritol Solu-tion. The notion is basically this—dumping billions of tons of iron solution into the ocean will stimulate massive plant growth that will suck enough carbon dioxide out of the atmosphere to counter the effects of all the CO_2 humans are releasing into the atmosphere by burning fossil fuels. A test of the theory in 1995 transformed a patch of ocean near the Galápagos Islands from sparkling blue to murky green overnight, as the iron triggered the growth of massive amounts of phytoplankton.)

Because iron is so important, most medical research has focused on populations who don't get enough iron. Some doctors and nu-tritionists have operated under the assumption that more iron can only be better. The food industry currently supplements everything from flour to breakfast cereal to baby formula with iron.

You know what they say about too much of a good thing?

Our relationship with iron is much more complex than it's been considered traditionally. It's essential—but it also provides a pro-verbial leg up to just about every biological threat to our lives. With very few exceptions in the form of a few bacteria that use other metals in its place, almost all life on earth needs iron to survive.

Parasites hunt us for our iron; cancer cells thrive on our iron. Finding, controlling, and using iron is the game of life. For bacteria, fungi, and protozoa, human blood and tissue are an iron gold mine. Add too much iron to the human system and you may just be loading up the buffet table.

IN 1952, EUGENE D. WEINBERG was a gifted microbial researcher with a healthy curiosity and a sick wife. Diagnosed with a mild infection, his wife was prescribed tetracycline, an antibiotic. Professor Weinberg wondered whether anything in her diet could interfere with the effectiveness of the antibiotic. We've only scratched the surface of our understanding of bacterial interactions today; in 1952, medical science had only scratched the surface of the scratch. Weinberg knew how little we knew, and he knew how unpredictable bacteria could be, so he wanted to test how the antibiotic would react to the presence or absence of specific chemicals that his wife was adding to her system by eating.

In his lab, at Indiana University, he directed his assistant to load up dozens of petri dishes with three compounds: tetracycline, bacteria, and a third organic or elemental nutrient, which varied from dish to dish. A few days later, one dish was so loaded with bacteria that Professor Weinberg's assistant assumed she had forgotten to add the antibiotic to that dish. She repeated the test for that nutrient and got the same result—massive bacteria growth. The nutrient in this sample was providing so much booster fuel to the bacteria that it effectively neutralized the antibiotic. You guessed it—it was iron.

Weinberg went on to prove that access to iron helps nearly all bacteria multiply almost unimpeded. From that point on, he dedicated his life's work to understanding the negative effect that the ingestion of excess iron can have on humans and the relationship other life-forms have to it.

Human iron regulation is a complex system that involves virtually every part of the body. A healthy adult usually has between three and four grams of iron in his or her body. Most of this iron is in the bloodstream within hemoglobin, distributing oxygen, but iron can also be found throughout the body. Given that iron is not only crucial to our survival but can be a potentially deadly liability, it shouldn't be surprising that we have iron-related defense mechanisms as well.

We're most vulnerable to infection where infection has a gateway to our bodies. In an adult without wounds or broken skin, that means our mouths, eyes, noses, ears, and genitals. And because infectious agents need iron to survive, all those openings have been declared iron no-fly-zones by our bodies. On top of that, those openings are patrolled by chelators—proteins that lock up iron molecules and prevent them from being used. Everything from tears to saliva to mucus—all the fluids found in those bodily entry points—are rich with chelators.

There's more to our iron defense system. When we're first beset by illness, our immune system kicks into high gear and fights back with what is called the *acute phase response*. The bloodstream is flooded with illness-fighting proteins, and, at the same time, iron is locked away to prevent biological invaders from using it against us. It's the biological equivalent of a prison lockdown—flood the halls with guards and secure the guns.

A similar response appears to occur when cells become cancerous and begin to spread without control. Cancer cells require iron to grow, so the body attempts to limit its availability. New pharmaceutical research is exploring ways to mimic this response by developing drugs to treat cancer and infections by limiting their access to iron.

Even some folk cures have regained respect as our understanding of bacteria's reliance on iron has grown. People used to cover wounds with egg-white-soaked straw to protect them from infection. It turns out that wasn't such a bad idea—preventing infection is what egg whites are made for. Egg shells are porous so that the chick embryo inside can "breathe." The problem with a porous shell, of course, is that air isn't the only thing that can get through it—so can all sorts of nasty microbes. The egg white's there to stop them. Egg whites are chock-full of chelators (those iron locking proteins that patrol our bodies' entry points) like ovoferrin in order to protect the developing chicken embryo—the yolk—from infection.

The relationship between iron and infection also explains one of the ways breast-feeding helps to prevent infections in newborns. Mother's milk contains lactoferrin—a chelating protein that binds with iron and prevents bacteria from feeding on it.

BEFORE WE RETURN to Aran Gordon and hemochromatosis, we need to take a side trip, this time to Europe in the middle of the fourteenth century—not the best time to visit.

From 1347 through the next few years, the bubonic plague swept across Europe, leaving death, death, and more death in its wake. Somewhere between one-third and one-half of the population was killed—more than 25 million people. No recorded pandemic, before or since, has come close to touching the plague's record. We hope none ever will.

It was a gruesome disease. In its most common form the bacterium that's thought to have caused the plague (*Yersinia pestis*, named after Alexander Yersin, one of the bacteriologists who first isolated it in 1894) finds a home in the body's lymphatic system, painfully swelling the lymph nodes in the armpits and groin until those swollen lymph nodes literally burst through the skin. Untreated, the survival rate is about one in three. (And that's just the bubonic form, which infects the lymphatic system; when *Y. pestis* makes it into the lungs and becomes airborne, it kills nine out of ten—and not only is it more lethal when it's airborne, it's more contagious!)

The most likely origin of the European outbreak is thought to be a fleet of Genoese trading ships that docked in Messina, Italy, in the fall of 1347. By the time the ships reached port, most of the crews were already dead or dying. Some of the ships never even made it to port, running aground along the coast after the last of their crew became too sick to steer the ship. Looters preyed on the wrecks and got a lot more than they bargained for—and so did just about everyone they encountered as they carried the plague to land.

In 1348 a Sicilian notary named Gabriele de'Mussi tells of how the disease spread from ships to the coastal populations and then inward across the continent:

> Alas! Our ships enter the port, but of a thousand sailors hardly ten are spared. We reach our homes; our kindred ... come

from all parts to visit us. Woe to us for we cast at them the
darts of death! ... Going back to their homes, they in turn
soon infected their whole families, who in three days suc-
cumbed, and were buried in one common grave.

Panic rose as the disease spread from town to town. Prayer
vigils were held, bonfires were lighted, churches were filled
with throngs. Inevitably, people looked for someone to blame.
First it was Jews, and then it was witches. But rounding them up
and burning them alive did nothing to stop the plague's deadly
march.

Interestingly, it's possible that practices related to the observance of Passover helped to protect Jewish neighborhoods from
the plague. Passover is a week-long holiday commemorating Jews'
escape from slavery in Egypt. As part of its observance, Jews do
not eat leavened bread and remove all traces of it from their homes.
In many parts of the world, especially Europe, wheat, grain, and
even legumes are also forbidden during Passover. Dr. Martin J.
Blaser, a professor of internal medicine at New York University
Medical Center, thinks this "spring cleaning" of grain stores may
have helped to protect Jews from the plague, by decreasing their
exposure to rats hunting for food—rats that carried the plague.

Victims and physicians alike had little idea what was causing
the disease. Communities were overwhelmed simply by the volume of bodies that needed burying. And that, of course, contrib-
uted to the spread of the disease as rats fed on infected corpses,
fleas fed on infected rats, and additional humans caught the disease from infected fleas. In 1348 a Sienese named Agnolo di Tura
wrote:

Father abandoned child, wife husband, one brother another, for this illness seemed to strike through the breath and sight. And so they died. And none could be found to bury the dead for money or friendship. Members of a household brought their dead to a ditch as best they could, without priest, without divine offices ... great pits were dug and piled deep with the multitude of dead. And they died by the hundreds both day and night. ... And as soon as those ditches were filled more were dug. ... And I, Agnolo di Tura, called the Fat, buried my five children with my own hands. And there were also those who were so sparsely covered with earth that the dogs dragged them forth and devoured many bodies throughout the city. There was no one who wept for any death, for all awaited death. And so many died that all believed it was the end of the world.

As it turned out, it wasn't the end of the world, and it didn't kill everyone on earth or even in Europe. It didn't even kill everyone it infected. Why? Why did some people die and others survive?

The emerging answer may be found in the same place Aran Gordon finally found the answer to his health problem—iron. New research indicates that the more iron in a given population, the more vulnerable that population is to the plague. In the past, healthy adult men were at greater risk than anybody else—children and the elderly tended to be malnourished, with corresponding iron deficiencies, and adult women are regularly iron depleted by menstruation, pregnancy, and breast-feeding. It might be that, as Stephen Ell, a professor at the University of Iowa, wrote, "Iron sta-

tus mirror[ed] mortality. Adult males were at highest risk on this
basis, with women [who lose iron through menstruation], children,
and the elderly relatively spared."

There aren't any highly reliable mortality records from the fourteenth century, but many scholars believe that men in their prime
were the most vulnerable. More recent—but still long ago—outbreaks of bubonic plague, for which there are reliable mortality records, demonstrate that the perception of heightened vulner-
ability in healthy adult men is very real. A study of plague in St.
Botolph's Parish in 1625 indicates that men between fifteen and
forty-four killed by the disease outnumbered women of the same
age by a factor of two to one.

SO LET'S GET back to hemochromatosis. With all this iron in
their systems, people with hemochromatosis should be magnets
for infection in general and the plague in particular, right?

Wrong.

Remember the iron-locking response of the body at the onset
of illness? It turns out that people who have hemochromatosis have
a form of iron locking going on as a permanent condition. The excess iron that the body takes on is distributed throughout the
body—but it isn't distributed *everywhere* throughout the body. And
while most cells end up with too much iron, one particular type of
cell ends up with much *less* iron than normal. The cells that hemochromatosis is stingy with when it comes to iron are a type of white
blood cell called *macrophages*. Macrophages are the police wagons
of the immune system. They circle our systems looking for trouble;
when they find it, they surround it, try to subdue or kill it, and
bring it back to the station in our lymph nodes.

In a nonhemochromatic person, macrophages have plenty of iron. Many infectious agents, like tuberculosis, can use that iron within the microphage to feed and multiply (which is exactly what the body is trying to prevent through the iron-locking response). So when a normal macrophage gathers up certain infectious agents to protect the body, it inadvertently is giving those infectious agents a Trojan horse access to the iron they need to grow stronger. By the time those macrophages get to the lymph node, the invaders in the wagon are armed and dangerous and can use the lymphatic system to travel throughout the body. That's exactly what happens with bubonic plague: the swollen and bursting lymph nodes that characterize it are the direct result of the bacteria's subversion of the body's immune system for its own purposes.

Ultimately, the ability to access iron within our macrophages is what makes some intracellular infections deadly and others benign. The longer our immune system is able to prevent an infection from spreading by containing it, the better it can develop other means, like antibodies, to overwhelm it. If your macrophages lack iron, as they do in people who have hemochromatosis, those macrophages have an additional advantage—not only do they isolate infectious agents and cordon them off from the rest of the body, they also starve those infectious agents to death.

New research has demonstrated that iron-deficient macrophages are indeed the Bruce Lees of the immune system. In one set of experiments, macrophages from people who had hemochromatosis and macrophages from people who did not were matched against bacteria in separate dishes to test their killing ability. The hemochromatic macrophages crushed the bacteria— they are thought to be significantly better at combating bacteria by

limiting the availability of iron than the nonhemochromatic macrophages.

Which brings us full circle. Why would you take a pill that was
guaranteed to kill you in forty years? Because it will save you tomorrow. Why would we select for a gene that will kill us through
iron loading by the time we reach what is now middle age? Because
it will protect us from a disease that is killing everyone else long
before that.

HEMOCHROMATOSIS IS CAUSED by a genetic mutation. It pre-
dates the plague, of course. Recent research has suggested that it
originated with the Vikings and was spread throughout Northern
Europe as the Vikings colonized the European coastline. It may
have originally evolved as a mechanism to minimize iron deficiencies in poorly nourished populations living in harsh environments.
(If this was the case, you'd expect to find hemochromatosis in all
populations living in iron-deficient environments, but you don't.)
Some researchers have speculated that women who had hemochromatosis might have benefited from the additional iron absorbed through their diet because it prevented anemia caused by
menstruation. This, in turn, led them to have more children, who
also carried the hemochromatosis mutation. Even more specula-
tive theories have suggested that Viking men may have offset the
negative effects of hemochromatosis because their warrior culture
resulted in frequent blood loss.

As the Vikings settled the European coast, the mutation may
have grown in frequency through what geneticists call the founder
effect. When small populations establish colonies in unpopulated

or secluded areas, there is significant inbreeding for generations. This inbreeding virtually guarantees that any mutations that aren't fatal at a very early age will be maintained in large portions of the population.

Then, in 1347, the plague begins its march across Europe. People who have the hemochromatosis mutation are especially resistant to infection because of their iron-starved macrophages. So, though it will kill them decades later, they are much more likely than people without hemochromatosis to survive the plague, reproduce, and pass the mutation on to their children. In a population where most people don't survive until middle age, a genetic trait that will kill you when you get there but increases your chance of arriving is—well, something to ask for.

The pandemic known as the Black Death is the most famous—and deadly—outbreak of bubonic plague, but historians and scientists believe there were recurring outbreaks in Europe virtually every generation until the eighteenth or nineteenth century. If hemochromatosis helped that first generation of carriers to survive the plague, multiplying its frequency across the population as a result, it's likely that these successive outbreaks compounded that effect, further breeding the mutation into the Northern and Western European populations every time the disease resurfaced over the ensuing three hundred years. The growing percentage of hemochromatosis carriers—potentially able to fend off the plague—may also explain why no subsequent epidemic was as deadly as the pandemic of 1347 to 1350.

This new understanding of hemochromatosis, infection, and iron has provoked a reevaluation of two long-established medical treatments—one very old and all but discredited, the other more recent and all but dogma. The first, bleeding, is back; the second, iron dosing, especially for anemics, is being reconsidered in many circumstances.

BLOODLETTING IS ONE of the oldest medical practices in history, and nothing has a longer or more complicated record. First recorded three thousand years ago in Egypt, it reached its peak in the nineteenth century only to be roundly discredited as almost savage over the last hundred years. There are records of Syrian doctors using leeches for bloodletting more than two thousand years ago and accounts of the great Jewish scholar Maimonides' employing bloodletting as the physician to the royal court of Saladin, sultan of Egypt, in the twelfth century. Doctors and shamans from Asia to Europe to the Americas used instruments as varied as sharpened sticks, shark's teeth, and miniature bows and arrows to bleed their patients.

In Western medicine, the practice was derived from the thinking of the Greek physician Galen, who practiced the theory of the four humours—blood, black bile, yellow bile, and phlegm. According to Galen and his intellectual descendants, all illness resulted from an imbalance of the four humours, and it was the doctor's job to balance those fluids through fasting, purging, and bloodletting.

Volumes of old medical texts are devoted to how and how much blood should be drawn. An illustration from a 1506 book on medicine points to forty-three different places on the human body that should be used for bleeding—fourteen on the head alone.

For centuries in the West, the place to go for bloodletting was the barber shop. In fact, the barber's pole originated as a symbol for bloodletting—the brass bowl at the top represented the bowl where leeches were kept; the one at the bottom represented the bowl for collecting blood. And the red and white spirals have their origins in the medieval practice of hanging bandages on a pole to dry them after they were washed. The bandages would twist in the wind and wrap themselves in spirals around the pole. As to why barbers were the surgeons of the day? Well, they were the guys with the razor blades.

Bloodletting reached its peak in the eighteenth and nineteenth centuries. According to medical texts of the time, if you presented to your doctor with a fever, hypertension, or dropsy, you would be bled. If you had an inflammation, apoplexy, or a nervous disorder, you would be bled. If you suffered from a cough, dizziness, headache, drunkenness, palsy, rheumatism, or shortness of breath, you would be bled. As crazy as it sounds, even if you were hemorrhaging blood you would be bled.

Modern medical science has been skeptical of bloodletting for many reasons—at least some of them deserved. First of all, eighteenth- and nineteenth-century reliance on bleeding as a treatment for just about everything is reasonably suspect.

When George Washington was ill with a throat infection, doctors treating him conducted at least four bleedings in just twenty-four hours. It's unclear today whether Washington actually died from the infection or from shock caused by blood loss. Doctors in the nineteenth century routinely bled patients until they fainted; they took that as a sign they'd removed just the right amount of blood.

After millennia of practice, bloodletting fell into extreme disfavor at the beginning of the twentieth century. The medical community—even the general public—considered bleeding to be the epitome of everything that was barbaric about prescientific medicine. Now, new research indicates that—like so much else—the broad discrediting of bloodletting may have been a rush to judgment.

First of all, it's now absolutely clear that bloodletting—or phlebotomy, as it's known today—is the treatment of choice for hemochromatosis patients. Regular bleeding of hemochromatosis patients reduces the iron in their systems to normal levels and prevents the iron buildup in the body's organs that is so damaging.

It's not just for hemochromatosis, either—doctors and researchers are examining phlebotomy as an aid in combating heart disease, high blood pressure, and pulmonary edema. And even our complete dismissal of historic bloodletting practices is getting another look. New evidence suggests that, in moderation, bloodletting may have had a beneficial effect.

A Canadian physiologist named Norman Kasting discovered that bleeding animals induces the release of the hormone vasopressin; this reduces their fevers and spurs their immune system into higher gear. The connection isn't unequivocally proven in humans, but there is much correlation between bloodletting and fever reduction in the historic record. Bleeding also may have helped to fight infection by reducing the amount of iron available to feed an invader, providing an assist to the body's natural tendency to hide iron when it recognizes an infection.

When you think about it, the notion that humans across the globe continued to practice phlebotomy for thousands of years probably indicates that it produced *some* positive results. If everyone who was treated with bloodletting died, its practitioners would have been out of business pretty quickly.

One thing is clear—an ancient medical practice that "modern" medical science dismissed out of hand is the only effective treatment for a disease that would otherwise destroy the lives of thousands of people. The lesson for medical science is a simple one—there is much more that the scientific community doesn't understand than there is that it does understand.

IRON IS GOOD. Iron is good. Iron is good.

Well, now you know that, like just about every other good thing under the sun, when it comes to iron, it's moderation, moderation, moderation. But until recently, current medical thinking didn't recognize that. Iron was thought to be good, so the more iron the better.

A doctor named John Murray was working with his wife in a Somali refugee camp when he noticed that many of the nomads, despite pervasive anemia and repeated exposure to a range of virulent pathogens, including malaria, tuberculosis, and brucellosis, were free of visible infection. He responded to this anomaly by deciding to treat only part of the population with iron at first. Sure enough, he treated some of the nomads for anemia by giving them iron supplements, and suddenly the infections gained the upper hand. The rate of infection in nomads receiving the extra iron skyrocketed. The Somali nomads weren't withstanding these infections *despite* their anemia: they were withstanding these infections *because of* their anemia. It was iron locking in high gear.

Thirty-five years ago, doctors in New Zealand routinely injected Maori babies with iron supplements. They assumed that the Maori (the indigenous people of New Zealand) had a poor diet, lacking iron, and that their babies would be anemic as a result.

The Maori babies injected with iron were seven times as likely to suffer from potentially deadly infections, including septicemias (blood poisoning) and meningitis. Like all of us, babies have isolated strains of potentially harmful bacteria in their systems, but those strains are normally kept under control by their bodies. When the doctors gave these babies iron boosters, they were giving booster fuel to the bacteria, with tragic results.

It's not just iron dosing through injection that can cause this blossoming of infections; iron-supplemented food can be food for bacteria too. Many infants can have botulism spores in their intestines (the spores can be found in honey, and that's one of the reasons parents are warned not to feed honey to babies, especially before they turn one). If the spores germinate, the results can be fatal. A study of sixty-nine cases of infant botulism in California showed one key difference between fatal and nonfatal cases of botulism in babies. Babies who were fed with iron-supplemented formula instead of breast-fed were much younger when they began to get sick and more vulnerable as a result. Of the ten who died, all had been fed with the iron-enhanced formula.

By the way, hemochromatosis and anemia aren't the only hereditary diseases that have gained pride of place in our gene pool by offering protection from another threat, and they're not all related to iron. The second most common genetic disease in Europeans, after hemochromatosis, is cystic fibrosis. It's a terrible, debilitating disease that affects different parts of the body. Most people with cystic fibrosis die young, usually from lung-related illness. Cystic fibrosis is caused by a mutation in a gene called CFTR; it takes two copies of the mutated gene to cause the disease. Somebody with only one copy of the mutated gene is known as a carrier but does not have cystic fibrosis. It's thought that at least 2 percent of people descended from Europeans are carriers, making the mutation very common indeed from a genetic perspective. New research suggests that, sure enough, carrying a copy of the gene that causes cystic fibrosis seems to offer some protection from tuberculosis. Tuberculosis, which has also been called consumption because of the way it seems to consume its victims from the inside out, caused 20 percent of all the deaths in Europe between 1600 and 1900, making it a very deadly disease. And making anything that helped to protect people from it look pretty attractive while lounging in the gene pool.

ARAN GORDON FIRST manifested symptoms of hemochromatosis as he began training for the Marathon des Sables—that grueling 150-mile race across the Sahara Desert. But it would take three years of progressive health problems, frustrating tests, and inaccurate conclusions before he finally learned what was wrong with him. When he did, he was told that untreated he had five years to live.

Today, we know that Aran suffered the effects of the most common genetic disorder in people of European descent—hemochromatosis, a disorder that may very well have helped his ancestors to survive the plague.

Today, Aran's health has been restored through bloodletting, one of the oldest medical practices on earth.

Today, we understand much more about the complex interrelationship of our bodies, iron, infection, and conditions like hemochromatosis and anemia.

What doesn't kill us, makes us stronger.

Which is probably some version of what Aran Gordon was thinking when he finished the Marathon des Sables for the second time in April 2006—just a few months after he was supposed to have died.

CHAPTER 11

A SPOONFUL OF SUGAR HELPS THE TEMPERATURE GO DOWN

The World Health Organization estimates that 171 million people have diabetes—and that number is expected to double by 2030. You almost certainly know people with diabetes—and you certainly have heard of people with diabetes. Halle Berry, Mikhail Gorbachev, and George Lucas all have diabetes. It's one of the most common chronic diseases in the world, and it's getting more common every day.

Diabetes is all about the body's relationship to sugar, specifically the blood sugar known as glucose. Glucose is produced when the body breaks down carbohydrates in the food we eat. It's essential to survival—it provides fuel for the brain; it's required to manufacture proteins; it's what we use to make energy when we need it.

With the help of insulin, a hormone made by the pancreas, glucose is stored in your liver, muscles, and fat cells (think of them as
your own internal OPEC) waiting to be converted to fuel as necessary.

The full name of the disease is actually *diabetes mellitus*—which
literally means "passing through honey sweet." One of the first
outward manifestations of diabetes is the need to pass large
amounts of sugary urine. And for thousands of years, observers
have noticed that diabetics' urine smells (and tastes) particularly
sweet. In the past Chinese physicians actually diagnosed and mon-
itored diabetes by looking to see whether ants were attracted to
someone's urine. In diabetics, the process through which insulin
helps the body use glucose is broken, and the sugar in the blood
builds up to dangerously high levels. Unmanaged, these abnormal
blood sugar levels can lead to rapid dehydration, coma, and death.
Even when diabetes is tightly managed, its long-term complications include blindness, heart disease, stroke, and vascular disease
that often leads to gangrene and amputation.

There are two major types of diabetes, Type 1 and Type 2, commonly called juvenile diabetes and adult-onset diabetes, respec-
tively, because of the age at which each type is usually diagnosed.
(Increasingly, adult-onset diabetes is becoming a misnomer: sky-
rocketing rates of childhood obesity are leading to increasing numbers of children who have Type 2 diabetes.)

Some researchers believe that Type 1 diabetes is an autoimmune disease—the body's natural defense system incorrectly iden-
tifies certain cells as outside invaders and sets out to destroy them.
In the case of Type 1 diabetes, the cells that fall victim to this biological friendly fire are the precise cells in the pancreas responsible
for insulin production. No insulin means the body's blood sugar

refinery is effectively shut down. As of today, Type 1 diabetes can only be treated with daily doses of insulin, typically through self-administered injections, although it is also possible to have an insulin pump surgically implanted. On top of daily insulin doses, Type 1 requires vigilant attention to blood sugar levels and a superdisciplined approach to diet and exercise.

In Type 2 diabetes, the pancreas still produces insulin—sometimes even at high levels—but the level of insulin production can eventually be too low or other tissues in the body are resistant to it, impairing the absorption and conversion of blood sugar. Because the body is still producing insulin, Type 2 diabetes can often be managed without insulin injections, through a combination of other medications, careful diet, exercise, weight loss, and blood sugar monitoring.

There is also a third type of diabetes, called gestational diabetes because it occurs in pregnant women. Gestational diabetes can be a temporary type of diabetes that tends to resolve itself after pregnancy. In the United States, it occurs in as much as 4 percent of pregnant women—some 100,000 expectant mothers a year. It can also lead to a condition in the newborn called macrosomia—which is a fancy term for "really chubby baby" as all the extra sugar in the mother's bloodstream makes its way across the placenta and feeds the fetus. Some researchers think this type of diabetes may be "intentionally" triggered by a hungry fetus looking for Mommy to stock the buffet table with sugary glucose.

So what causes diabetes? The truth is, we don't fully understand. It's a complex combination that can involve inheritance, infections, diet, and environmental factors. At the very least, inheritance definitely causes a predisposition to diabetes that can be triggered by some other factor. In the case of Type 1 diabetes, that trigger may be a virus or even an environmental trigger. In the case of Type 2, scientists think many people pull the trigger themselves through poor eating habits, lack of exercise, and resulting obesity. But one thing is clear—genetics contributes to Type 1 and especially to Type 2 diabetes. And that's where, for our purposes, things really start to heat up. Or, more precisely, to cool down, as you'll see shortly.

THERE'S A BIG difference in the prevalence of Type 1 and Type 2 diabetes that is largely based on geographic origin. Even though there seems to be a stronger genetic component to Type 2 diabetes, it is also closely related to lifestyle; 85 percent of people who have this type of diabetes are obese. That means it's currently much more common in the developed world because easy access to high-calorie, low-nutrient junk food means so many more people are obese—but it seems clear that the predisposition to Type 2 diabetes exists across population groups. There are higher levels of incidence in certain populations, of course—but even that tends to occur hand in hand with higher levels of obesity. The Pima Indians of the southwestern United States, for example, have a staggering rate of diabetes—nearly half of all adults. It's possible that their historic hunter-gatherer lifestyle produced metabolisms more suited for the Atkins diet than the carbohydrate- and sugar-heavy diet that European farmers survived on for centuries. Type 1 diabetes is different—it is much, much more common in people of Northern European descent. Finland has the highest rate of juvenile diabetes in the world. Sweden is second, and the United Kingdom and Norway are tied for third. As you head south, the rate drops lower and lower. It's downright uncommon in people of purely African, Asian, and Hispanic descent.

When a disease that is caused at least partially by genetics is significantly more likely to occur in a specific population, it's time to raise the evolutionary eyebrows and start asking questions— because that almost certainly means that some aspect of the trait that causes the disease today helped the forebears of that population group to survive somewhere back up the evolutionary line.

In the case of hemochromatosis, we know that the disease probably provided carriers with protection from the plague by denying the bacteria that causes it the iron it needs to survive. So what could diabetes possibly do for us? To answer that, we're going to take another trip down memory lane—this time measured, not in centuries, but in millennia. Put your ski jackets on; we're looking for an ice age.

UNTIL ABOUT FIFTY years ago, the conventional wisdom among scientists who studied global climate change was that large-scale climate change occurred very slowly. Today, of course, people from Al Gore to Julia Roberts are on a mission to make it clear that humanity has the power to cause cataclysmic change in just a few generations. But before the 1950s, most scientists believed that climate change took thousands, probably hundreds of thousands, of years.

That doesn't mean they didn't accept the notion that glaciers and ice sheets had once covered the Northern Hemisphere. They were just happily certain that glaciers moved, well, glacially: eons to descend and epochs to recede. Humanity certainly didn't have to worry about it—nobody was ever going to be run over by a speeding glacier. If massive climate change was going to lead us into a new ice age, we'd have a few hundred thousand years to do something about it.

Of course, there were some contrary voices singing a different tune, but the larger scientific community paid them very little regard. Andrew Ellicott Douglass was an astronomer working in Arizona in 1895 when he first started cutting down trees to examine them for evidence of any effect from a specific solar activity, called sunspots, that occurs in cycles. He never found that—but he did ultimately invent dendrochronology, the scientific technique of studying tree rings for clues about the past. One of his first observations was that tree rings were thinner during cold or dry years and thicker during wet or warm years. And by rolling back the years, one ring at a time, he discovered what appeared to be a century-long climate change that occurred around the seventeenth century, with a significant drop in temperature. The reaction of the scientific community was a collective "Nah." As far as the climate change community was concerned, Douglass was cutting down trees in a forest with nobody there to hear it. (According to Dr. Lloyd Burckle of Columbia University, not only was Douglass right: the hundred-year cold spell he discovered was responsible for some beautiful music. Burckle says the superior sound of the great European violin makers, including the famous Stradivari, is the result of the high-density wood from the trees that grew during this century-long freeze—denser because they grew less during the cold and had thinner rings as a result.)

More evidence of the possibility of rapid climate change accumulated. In Sweden, scientists studying layers of mud from lake bottoms found evidence of climate change that occurred much more quickly than anyone at the time thought possible. These scientists discovered large amounts of pollen from an Arctic wildflower called *Dryas octopetala* in mud cores from only 12,000 years ago. Dryas's usual home is the Arctic; it only truly flourished across Europe during periods of significant cold. Its widespread prevalence in Sweden around 12,000 years ago seemed to indicate that the warm weather that had followed the last ice age had been interrupted by a rapid shift back to much colder weather. In honor of the telltale wildflower, they named this arctic reprise the Younger Dryas. Of course, given prevailing thinking, even these scientists believed that the "rapid" onset of the Younger Dryas took 1,000 years or so.

It's hard to underestimate the chilling effect conventional wisdom can have on the scientific community. Geologists of the time believed the present was the key to the past—if this is the way the climate behaves today, that's the way it behaved yesterday. That philosophy is called uniformitarianism and, as the physicist Spencer Weart points out in his 2003 book *The Discovery of Global Warming*, it was the guiding principle among scientists of the time:

> Through most of the 20th century, the uniformitarian principle was cherished by geologists as the very foundation of their science. In human experience, temperatures apparently did not rise or fall radically in less than millennia, so the uniformitarian principle declared that such changes had never happened in the past.

If you're positive something doesn't exist, you're not going to look for it, right? And because everyone was certain that global climate changes took at least a thousand years, nobody even bothered to look at the evidence in a way that could reveal faster change. Those Swedish scientists studying the layers of lake bottom clay who first postulated the "rapid" thousand-year onset of the Younger Dryas? They were looking at chunks of mud spanning centuries; they never looked at samples small enough to demonstrate faster change. The proof that the Younger Dryas descended on the Northern Hemisphere much more rapidly than they thought was right in front of their eyes—but they were blinded by their assumptions.

BY THE 1950S and 1960s, the uniformitarian vise started to lose its hold, or at least change its grip, as scientists began to understand the potential of catastrophic events to produce rapid change. In the late 1950s, Dave Fultz at the University of Chicago built a mock-up of the earth's atmosphere using rotating fluids that simulated the behavior of atmospheric gases. Sure enough, the fluids moved in stable, repeating patterns—unless, that is, they were disturbed. Then, even the smallest interference could produce massive changes in the currents. It wasn't proof by a long shot, but it certainly was a powerful suggestion that the real atmosphere was susceptible to significant change. Other scientists developed mathematical models that indicated similar possibilities for rapid shifts.

As new evidence was discovered and old evidence was reexamined, the scientific consensus evolved. By the 1970s there was general agreement that the temperature shifts and climate changes leading into and out of ice ages could occur over mere hundreds of years. Thousands were out, hundreds were in. Centuries were the new "rapid."

There was a new consensus around when—but a total lack of agreement about how. Perhaps methane bubbled up from tundra bogs and trapped the heat of the sun. Perhaps ice sheets broke off from the Antarctic and cooled the oceans. Maybe a glacier melted into the North Atlantic, creating a massive freshwater lake that suddenly interrupted the ocean's delivery of warm tropical water to the north.

It's fitting that hard, cold proof was eventually found in hard, cold ice.

In the early 1970s, climatologists discovered that some of the best records of historic weather patterns were filed away in the glaciers and ice plateaus of northern Greenland. It was hard, treacherous work—if you're imagining the stereotypical lab rat in a white coat, think again. This was *Extreme Sports: Ph.D.*—multinational teams trekking across miles of ice, climbing thousands of feet, hauling tons of machines, and enduring altitude sickness and freakish cold, all so they could bore into a two-mile core of ice. But the prize was a pristine and unambiguous record of yearly precipitation and past temperature, unspoiled by millennia and willing to reveal its secrets with just a little chemical analysis. Once you paid it a visit, of course.

By the 1980s, these ice cores definitively confirmed the existence of the Younger Dryas—a severe drop in temperature that began around 13,000 years ago and lasted more than a thousand years. But that was just, well, the tip of the iceberg.

In 1989 the United States mounted an expedition to drill a core all the way to the bottom of the two-mile Greenland ice sheet—representing 110,000 years of climate history. Just twenty miles away, a European team was conducting a similar study. Four years later, both teams got to the bottom—and the meaning of *rapid* was about to change again.

The ice cores revealed that the Younger Dryas—the last ice age—ended in just three years. Ice age to no ice age—not in three thousand years, not in three hundred years, but in three plain years. What's more, the ice cores revealed that the onset of the Younger Dryas took just a decade. The proof was crystal clear this time—rapid climate change was very real. It was so rapid that scientists stopped using the word *rapid* to describe it, and started using words like *abrupt* and *violent*. Dr. Weart summed it up in his 2003 book:

> Swings of temperature that scientists in the 1950s believed to take tens of thousands of years, in the 1970s to take thousands of years, and in the 1980s to take hundreds of years, were now found to take only decades.

In fact, there have been around a score of these abrupt climate changes over the last 110,000 years; the *only* truly stable period has been the last 11,000 years or so. Turns out, the present isn't the key to the past—it's the exception.

The most likely suspect for the onset of the Younger Dryas and the sudden return to ice age temperatures across Europe is the breakdown of the ocean "conveyor belt," or thermohaline circulation, in the Atlantic Ocean. When it's working normally—or at least the way we're used to it—the conveyor carries warm tropical water on the ocean surface to the north, where it cools, becomes denser, sinks, and is carried south through the ocean depths back to the Tropics. Under those circumstances, Britain is temperate even though it's on the same latitude as much of Siberia. But when the conveyor is disrupted—say, by a huge influx of warm fresh water melting off the Greenland ice sheet—it may have a significant impact on global climate and turn Europe into a very, very cold place.

JUST BEFORE THE Younger Dryas, our European ancestors were doing pretty well. Tracing human migration through DNA, scientists have documented a population explosion in Northern Europe as populations that had once migrated north out of Africa now moved north again into areas of Europe that had been uninhabitable during the last ice age (before the Younger Dryas). The average temperature was nearly as warm as it is today, grasslands flourished where glaciers had once stood, and human beings thrived.

And then the warming trend that had persisted since the end of the last ice age kicked rapidly into reverse. In just a decade or so, average yearly temperatures plunged nearly thirty degrees. Sea levels dropped by hundreds of feet as water froze and stayed in the ice caps. Forests and grasslands went into a steep decline. Coastlines were surrounded by hundreds of miles of ice. Icebergs were common as far south as Spain and Portugal. The great, mountainous glaciers marched south again. The Younger Dryas had arrived, and the world was changed.

Though humanity would survive, the short-term impact, especially for those populations that had moved north, was devastating. In less than a generation, virtually every learned method of survival—from the shelters they built to the hunting they practiced—was inadequate. Many thousands of humans almost certainly froze or starved to death. Radiocarbon dating from archaeological sites provides clear evidence that the human population in Northern

Europe went into a steep decline, showing a steep drop-off in set-tlements and other human activity.

But humans clearly survived; the question is, how? Certainly some of our success was due to social adaptation—many scientists think that the Younger Dryas helped to spur the collapse of hunter-gatherer societies and the first development of agriculture. But what about biological adaptation and natural selection? Scientists believe some animals perfected their natural ability to survive cold spells during this period—notably the wood frog, which we'll re-turn to later. So why not humans? Just as the European population may have "selected" for the hemochromatosis gene because it helped its carriers withstand the plague, might some other genetic trait have provided its carriers with superior ability to withstand the cold? To answer that, let's take a look at the effect of cold on humans.

IMMEDIATELY UPON HIS death in July 2002, baseball legend Ted Williams was flown to a spa in Scottsdale, Arizona, checked in, and given a haircut, a shave, and a cold plunge. Of course, this wasn't your typical Arizona spa—this was the Alcor Life Exten-sion cryonics lab, and Williams was checking in for the foreseeable future. According to his son, he hoped that future medical science might be able to restore him to life.

Alcor separated Williams's head from his body, drilled a couple of dime-size holes in it, and froze it in a bucket of liquid nitrogen at minus 320 degrees Fahrenheit. (His body got its own cold storage container.) Alcor brochures suggest that "mature nanotechnology" might be able to reanimate frozen bodies "perhaps by the mid-

21st century," but they also note that cryonics is a "last-in-first-out process wherein the first-in may have to wait a very long time."

Make that a very, very long time, like . . . never. Unfortunately for Williams and the other sixty-six superchilled cadavers at Alcor, human tissue doesn't react well to freezing. When water is frozen, it expands into sharp little crystals. When humans are frozen, the water in our blood freezes, and the ice shards cut blood cells and cause capillaries to burst. It's not dissimilar to the way a pipe bursts when the water's left on in an unheated house—except no repairman can fix it.

Of course, just because we can't survive a true deep freeze doesn't mean our bodies haven't evolved many ways to manage the cold. They have. Not only is your body keenly aware of the danger cold poses, it's got a whole arsenal of natural defenses. Think back to some time when you were absolutely freezing—standing still for hours on a frigid winter morning watching a parade, riding a ski lift with the wind whipping across the mountain. You start to shiver. That's your body's first move. When you shiver, the increased muscle activity burns the sugar stored in your muscles and creates heat. What happens next is less obvious, but you've felt the effect. Remember the uncomfortable combination of tingling and numbness in your fingers and toes? That's your body's next move.

As soon as the body senses cold, it constricts the thin web of capillaries in your extremities, first your fingers and toes, then farther up your arms and legs. As your capillary walls close in, blood is squeezed out and driven toward your torso, where it essentially provides a warm bath for your vital organs, keeping them at a safe temperature, even if it means the risk of frostbite for your extremities. It's natural triage—lose the finger, spare the liver.

In people whose ancestors lived in particularly cold climates—like Norwegian fishermen or Inuit hunters—this autonomic response to cold has evolved with a further refinement. After some time in the cold, the constricted capillaries in your hands will dilate briefly, sending a rush of warm blood into your numbed fingers and toes before constricting again to drive the blood back into your core. This intermittent cycle of constriction and release is called the Lewis wave or "hunter's response," and it can provide enough warmth to protect your extremities from real injury, while still ensuring that your vital organs are safe and warm. Inuit hunters can raise the temperature in the skin of their hands from near freezing to fifty degrees in a mater of minutes; for most people it takes much longer. On the other hand, people descended from warm-weather populations don't seem to have this natural ability to protect their limbs and their core at the same time. During the frigid cold of the Korean War, African American soldiers were much more prone to frostbite than other soldiers.

Shivering and blood vessel constriction aren't the only ways the body generates and preserves heat. A portion of the fat in newborns and some adults is specialized heat-generating tissue called brown fat, which is activated when the body is exposed to cold. When blood sugar is delivered to a brown fat cell, instead of being stored for future energy as it is in a regular fat cell, the brown fat cell converts it to heat on the spot. (For someone acclimated to very cold temperatures, brown fat can burn up to 70 percent more fat.) Scientists call the brown fat process nonshivering thermogenesis, because it's heat creation without muscle movement. Shivering, of course, is only good for a few hours; once you exhaust the blood sugar stores in your muscles and fatigue sets in, it doesn't work anymore. Brown fat, on the other hand, can go on generating heat for as long as it's fed, and unlike most other tissues, it doesn't need insulin to bring sugar into cells.

Nobody's written the *Brown Fat Diet Book* yet because it requires more than your usual lifestyle change. Adults who don't live in extreme cold don't really have much, if any, brown fat. To accumulate brown fat and get it really working, you need to live in extreme cold for a few weeks. We're talking North Pole cold. And that's not all—you've got to stay there. Once you stop sleeping in your igloo, your brown fat stops working.

The body has one more response to the cold that's not completely understood—but you've probably experienced it. When most people are exposed to cold for a while, they need to pee. This response has puzzled medical researchers for hundreds of years. It was first noted by one Dr. Sutherland, in 1764, who was trying to document the benefits of submersing patients in the supposedly healing—but cold—waters of Bath and Bristol, England. After immersing a patient who suffered from "dropsy, jaundice, palsy, rheumatism and inveterate pain in his back," Sutherland noted that the patient was "pissing more than he drank." Sutherland chalked the reaction up to external water pressure, figuring (quite wrongly) that fluid was simply being squeezed out of his patient, and it wasn't until 1909 that researchers connected increased urine flow, or diuresis, to cold exposure.

The leading explanation for cold diuresis—the need to pee when it's cold—is still pressure; but not external pressure, internal pressure. The theory is that as blood pressure climbs in the body's core because of constriction in the extremities, the body signals the kidneys to offload some of the extra fluid. But that theory doesn't fully explain the phenomenon, especially in light of recent studies.

The U.S. Army Research Institute of Environmental Medicine

has conducted more than twenty years of study into human response to extreme heat, cold, depth, and altitude. Their research conclusively demonstrates that even highly cold-acclimated individuals still experience cold diuresis when the temperature dips toward freezing. So the question persists: Why do we need to pee when we're cold? This certainly isn't the most pressing question facing medical researchers today—but as you'll soon discover, the possibilities are intriguing. And the answers may shed light on much bigger issues—like a disease that currently affects 171 million people.

LET'S PUT ASIDE the delicate subject of cold diuresis and turn to one much more suitable for the dinner table—ice wine: delicious, prized, and—supposedly—created by accident. Four hundred years ago, a German vintner was hoping to squeak just a few more growing days out of the late autumn when his fields were hit by a sudden frost, or so the story goes. The grapes were curiously shrunken, but, not wanting to let his entire harvest go to waste, he decided to pick the frozen grapes anyway and see what would come of it, hoping for the best. He let the grapes defrost and then pressed the crop as he usually did but was disappointed when it yielded just one-eighth of the juice he was expecting. Since he had nothing to lose, he put his meager yield through the fermentation process.

And discovered that he had a hit on his hands. The finished wine was insanely sweet. Since its first, semilegendary, certainly accidental harvest, some winemakers have specialized in ice wine, waiting every year for the first frost so they can harvest crops of frozen grapes. Among the many ways wine is rated, graded, and weighted today, it is measured on a "sugar scale." Typical table wine runs from 0 to 3 on the sugar scale. Ice wine runs from 18 to 28.

The shrunken nature of the grapes is due to water loss. Chemically speaking, it's not difficult to guess why grapes might have evolved to offload water at the onset of a freeze—the less water in the grape, the fewer ice crystals there are to damage the delicate membranes of the fruit.

How about the sharp increase in sugar concentration? That makes sense too. Ice crystals are only made of pure water—but the temperature at which they start to form depends on what else is suspended in the fluid where the water is found. Anything dissolved in water interferes with its ability to form the hexagonal latticework of solid ice crystals. Average seawater, for example, full of salt, freezes at around 28 degrees Fahrenheit instead of the 32 degrees we think of as water's freezing point. Think about the bottle of vodka some people keep in their freezer. Usually, alcohol is about 40 percent of the liquid volume in the bottle; it does a great job of interfering with the creation of ice—vodka doesn't freeze until you cool it down to around minus 20 degrees Fahrenheit. Even most water in nature doesn't freeze at exactly 32 degrees, because it usually contains trace minerals or other impurities that lower the freezing point.

Like alcohol, sugar is a natural antifreeze. The higher the sugar content in a liquid, the lower the freezing point. (Nobody knows more about sugar and freezing than the food service chemists at 7-Eleven who were in charge of developing a sugar-free Slurpee beverage. In regular Slurpees, the sugar is what helps to keep the frozen treat slurpable—it prevents the liquid from completely freezing. So when they tried to make sugar-free Slurpees, they kept making sugar-free blocks of ice. According to a company press release, it took two decades for researchers to develop a diet Slurpee by combining artificial sweeteners with undigestible sugar alcohols.) So when the grape dumps water at the first sign of frost, it's actually protecting itself in two ways—first, by reducing water volume; and second, by raising the sugar concentration of the water that remains. And that allows the grape to withstand colder temperature without freezing.

Eliminating water to deal with the cold? That sounds an awful lot like cold diuresis—peeing when you're cold. And higher levels of sugar? Well, we know where we've heard that; but before we get back to diabetes, let's make one more stop: the animal kingdom.

MANY ANIMALS THRIVE in the cold. Some amphibians, like the bullfrog, spend the winter in the frigid but unfrozen water at the bottom of lakes and rivers. The mammoth Antarctic cod happily swims beneath the Antarctic ice; its blood contains an antifreeze protein that sticks to ice crystals and prevents them from growing. On the Antarctic surface, the woolly bear caterpillar lives through temperatures as low as minus 60 degrees Fahrenheit for fourteen years, until it turns into a moth and flies off into the sunset for a few short weeks.

But of all the adaptations to cold under the sun—or hidden from it—none is as remarkable as the little wood frog's.

The wood frog, *Rana sylvatica,* is a cute little critter about two inches long with a dark mask across its eyes like Zorro's that lives across North America, from northern Georgia all the way up to Alaska, including north of the Arctic Circle. On early spring nights you can hear its mating call—a "brack, brack" that sounds something like a baby duck's. But until winter ends, you won't hear the wood frog at all. Like some animals, the wood frog spends the entire winter unconscious. But unlike hibernating mammals that go into a deep sleep, kept warm and nourished by a thick layer of insulating fat, the wood frog gives in to the cold entirely. It buries itself under an inch or two of twigs and leaves and then pulls a trick that—despite Ted Williams's possible hopes and Alcor's best efforts—seems to come straight out of a science fiction movie.

It freezes solid.

If you were on a winter hike and accidentally kicked one of these frogsicles out into the open, you'd undoubtedly assume it was dead. When completely frozen, it might as well be in suspended animation—it has no heartbeat, no breathing, and no measurable brain activity. Its eyes are open, rigid, and unnervingly white.

But if you pitched a tent and waited for spring, you'd eventually discover that little old *Rana sylvatica* has a few tricks up its frog sleeves. Just a few minutes after rising temperatures thaw the frog, its heartbeat miraculously sparks into gear and it gulps for air. It will blink a few times as color returns to its eyes, stretch its legs, and pull itself up into a sitting position. Not long after that, it will hop off, none the worse for wear, and join the chorus of defrosted frogs looking for a mate.

NOBODY KNOWS THE wood frog better than the brilliant and irrepressible Ken Storey, a biochemist from Ottawa, Canada, who, along with his wife, Janet, has been studying them since the early 1980s. Storey had been studying insects with the ability to tolerate freezing when a colleague told him about the wood frog's remarkable ability. His colleague had been collecting frogs for study and accidentally left them in the trunk of his car. Overnight, there was an unexpected frost and he awoke to discover a bag of frozen frogs. Imagine his surprise later that day when they thawed out on his lab table and started jumping around!

Storey was immediately intrigued. He was interested in cryopreservation—freezing living tissue to preserve it. Despite the bad rap it gets for its association with high-priced attempts to freeze the rich and eccentric for future cures, cryopreservation is a critical area of medical research that has the potential to yield many important advances. It has already revolutionized reproductive medicine by giving people the opportunity to freeze and preserve eggs and sperm.

The next step—the ability to extend the viability of large human organs for transplants—would be a huge breakthrough that could save thousands of lives every year. Today, a human kidney can be preserved for just two days outside the human body, while a heart can last only a few hours. As a result, organ transplants are always a race against the clock, with very little time to find the best match and get the patient, organ, and surgeon into the same operating room. Every day in the United States, a dozen people die because the organ they need hasn't become available in time. If donated organs could be frozen and "banked" for later revival and transplant, the rates of successful transplants would almost surely climb significantly.

But currently it's impossible. We know how to use liquid nitrogen to lower the temperature of tissue at the blinding speed of 600 degrees per minute, but it isn't good enough. We have not figured out how to freeze large human organs and restore them to full viability. And, as was mentioned, we're nowhere near the ability to freeze and restore a whole person.

So when Storey heard about the freezing frog, he jumped at the opportunity to study it. Frogs have the same major organs as humans, so this new direction for his research could prove amazingly useful. With all our technological prowess, we can't freeze and restore a single major human organ—and here was an animal that naturally manages the complex chemical wizardry of freezing and restoring all its organs more or less simultaneously. After many years of study (and many muddy nights trudging through the woodlands of southern Canada on wood frog hunts), the Storeys have learned a good deal about the secrets behind *Rana sylvatica*'s death-defying freezing trick.

Here's what they've uncovered: Just a few minutes after the frog's skin senses that the temperature is dropping near freezing, it begins to move water out of its blood and organ cells, and, instead of urinating, it pools the water in its abdomen. At the same time, the frog's liver begins to dump massive (for a frog) amounts of glucose into its bloodstream, supplemented by the release of additional sugar alcohols, pushing its blood sugar level up a hundredfold. All this sugar significantly lowers the freezing point of whatever water remains in the frog's bloodstream, effectively turning it into a kind of sugary antifreeze.

There's still water throughout the frog's body, of course; it's just been forced into areas where ice crystals will cause the least damage and where the ice itself might even have a beneficial effect. When Storey dissects frozen frogs he finds flat sheets of ice sandwiched between the skin and muscle of the legs. There will also be a big chunk of ice in the abdominal cavity surrounding the frog's organs; the organs themselves are largely dehydrated and look wizened as raisins. In effect, the frog has carefully put its own organs on ice, not unlike adding ice to coolers containing human organs as they're readied for transport to transplant. Doctors remove an organ, place it into a plastic bag, and then place the bag in a cooler full of crushed ice so the organ is kept as cool as possible without actually being frozen or damaged.

There's water in the frog's blood, too, but the rich concentration of sugar not only lowers the freezing point, it also minimizes damage by forcing the ice crystals that eventually form into smaller, less jagged shapes that won't puncture or slash the walls of cells or capillaries. Even all of this doesn't prevent every bit of damage, but the frog has that covered, too. During the winter months of its frozen sleep, the frog produces a large volume of a clotting factor called fibrinogen that helps to repair whatever damage might have occurred during freezing.

ELIMINATING WATER AND driving up sugar levels to deal with the cold: Grapes do it. Now we know that frogs do it. Is it possible that some humans adapted to do it, too?

Is it a coincidence that the people most likely to have a genetic propensity for a disease characterized by exactly that (excessive elimination of water and high levels of blood sugar) are people descended from *exactly those places most ravaged by the sudden onset of an ice age* about 13,000 years ago?

As a theory, it's hotly controversial, but diabetes may have helped our European ancestors survive the sudden cold of the Younger Dryas.

As the Younger Dryas set in, any adaptation to manage the cold, no matter how disadvantageous in normal times, might have made the difference between making it to adulthood and dying young. If you had the hunter's response, for instance, you would have an advantage in gathering food, because you were less likely to develop frostbite.

Now imagine that some small group of people had a different response to the cold. Faced with year-round frigid temperatures, their insulin supply slowed, allowing their blood sugar to rise somewhat. As in the wood frog, this would have lowered the freezing point of their blood. They urinated frequently, to keep internal water levels low. (A recent U.S. Army study shows there is very little harm caused by dehydration in cold weather.) Suppose these people used their brown fat to burn that oversupply of sugar in their blood to create heat. Perhaps they even produced additional clotting factor to repair tissue damage caused by particularly deep cold snaps. It's not hard to imagine that these people might have had enough of an advantage over other humans, especially if, like the wood frog, the spike in sugar was only temporary, to make it more likely that they would survive long enough to reach reproductive age.

There are tantalizing bits of evidence to bolster the theory.

When rats are exposed to freezing temperatures, their bodies become resistant to their own insulin. Essentially, they become what we would call diabetic in response to the cold.

In areas with cold weather, more diabetics are diagnosed in colder months; in the Northern Hemisphere, that means more diabetics are diagnosed between November and February than between June and September.

Children are most often diagnosed with Type 1 diabetes when temperatures start to drop in late fall.

Fibrinogen, the clotting factor that repairs ice-damaged tissue in the wood frog, also mysteriously peaks in humans during winter months. (Researchers are taking note—that may mean that cold weather is an important, but underappreciated, risk factor for stroke.)

A study of 285,705 American veterans with diabetes measured seasonal differences in their blood sugar levels. Sure enough, the veterans' blood sugar levels climbed dramatically in the colder months and bottomed out during the summer. More telling, the contrast between summer and winter was even more pronounced in those who lived in colder climates, with greater differences in seasonal temperature. Diabetes, it seems, has some deep connection to the cold.

WE DON'T KNOW enough today to state with certainty that the predisposition to Type 1 or Type 2 diabetes is related to human cold response. But we do know that some genetic traits that are potentially harmful today clearly helped our ancestors to survive and reproduce (hemochromatosis and the plague, for example). So while it's tempting simply to question how a condition that can cause early death today could ever confer a benefit, that doesn't look at the whole picture.

Remember, evolution is amazing—but it isn't perfect. Just about every adaptation is a compromise of sorts, an improvement in some circumstances, a liability in others. A peacock's brilliant tail feathers make him more attractive to females—and attract more attention from predators. Human skeletal structure allows us to walk upright and gives us large skulls filled with big brains—and the combination means an infant's head can barely make it through its mother's birth canal. When natural selection goes to work, it doesn't favor adaptations that make a given plant or animal "better"—just whatever it takes for it to increase the chances for survival in its current environment. And when there's a sudden change in circumstances that threatens to wipe out a population—a new infectious disease, a new predator, or a new ice age—natural selection will make a beeline for any trait that improves the chance of survival.

"Are they kidding?" said one doctor when told of the diabetes theory by a reporter. "Type 1 diabetes would result in severe ketoacidosis and early death."

Sure—today.

But what if a temporary diabetes-like condition occurred in a person who had significant brown fat living in an ice age environment? Food would probably be limited, so dietary blood-sugar load would already be low, and brown fat would convert most of that to heat, so the ice age "diabetic's" blood sugar, even with less insulin, might never reach dangerous levels. Modern-day diabetics, on the other hand, with little or no brown fat, and little or no exposure to constant cold, have no use—and thus no outlet—for the sugar that accumulates in their blood. In fact, without enough insulin the body of a severe diabetic starves no matter how much he or she eats.

The Canadian Diabetes Association has helped to fund Ken Storey's study of the incredible freezing frog. It understands that just because we haven't definitively linked diabetes and the Younger Dryas doesn't mean we shouldn't explore biological solutions to high blood sugar found elsewhere in nature. Cold-tolerant animals like the wood frog exploit the antifreezing properties of high blood sugar to survive. Perhaps the mechanisms they use to manage the

complications of high blood sugar will help lead us to new treatments for diabetes. Plants and microbes adapted to extreme cold
might produce molecules that could do the same.

Instead of dismissing connections, we need to have the curiosity to pursue them. And in the case of diabetes, sugar, water, and
cold, there are clearly plenty of connections to pursue.

CHAPTER III

THE CHOLESTEROL ALSO RISES

Everybody knows that humanity's relationship with the sun is multifaceted. As we all learned in elementary school, almost the entire global ecology of our planet depends on sufficient sunlight—beginning with the production of oxygen by plants through photosynthesis, without which we wouldn't have food to eat or air to breathe. And as we all have learned more and more over the last couple of decades, too much sun can be a bad thing on a global level and an individual one, throwing our environment into chaos by causing drought or causing deadly skin cancer.

But most people don't know that the sun is just as important on an individual, biochemical level—and the relationship is just as two-sided. Natural sunlight simultaneously helps your body to create vitamin D and destroys your body's reserves of folic acid—both of which are essential to your health. To manage this can't-live-with-you-can't-live-without-you relationship, different populations have evolved a combination of adaptations that, together, help to protect folic acid and ensure sufficient vitamin D production.

VITAMIN D IS a critical component of human biochemistry, especially to ensure the growth of healthy bones in children and the maintenance of healthy bones in adults. It ensures that our blood has sufficient levels of calcium and phosphorus. New research is discovering that it's also crucial to the proper function of the heart, the nervous system, the clotting process, and the immune system.

Without enough vitamin D, adults are prone to osteoporosis and children are prone to a disease called *rickets* that results in improper bone growth and deformity. Vitamin D deficiencies have also been shown to play a role in the development of dozens of diseases—everything from many different cancers to diabetes, heart disease, arthritis, psoriasis, and mental illness. Once the link between vitamin D and rickets was established early in the twentieth century, American milk was fortified with vitamin D, all but eliminating the disease in America.

We don't have to rely on fortified milk for vitamin D, however. Unlike most vitamins, vitamin D can be made by the body itself. (Generally speaking, a vitamin is an organic compound that an animal needs to survive but can usually obtain only from outside the body.) We make vitamin D by converting something else that, like the sun, has been getting a bad rap lately, but is 100 percent necessary for survival—cholesterol.

Cholesterol is required to make and maintain cell membranes. It helps the brain to send messages and the immune system to protect us against cancer and other diseases. It's a key building block in the production of estrogen and testosterone and other hormones. And it is the essential component in our manufacture of vitamin D through a chemical process that is similar to photosynthesis in its dependence on the sun.

When we are exposed to the right kind of sunlight, our skin converts cholesterol to vitamin D. The sunlight necessary for this process is ultraviolet B, or UVB, which typically is strongest when the sun is more or less directly overhead—for a few hours every day beginning around noon. In parts of the world that are farther from the equator, very little UVB reaches the earth during winter months. Fortunately, the body is so efficient at making vitamin D that, as long as people get sufficient sun exposure and have enough cholesterol, we can usually accumulate enough vitamin D reserves to get us through the darker months.

By the way, the next time you get your cholesterol checked, make a note of the season. Because sunlight converts cholesterol to vitamin D, cholesterol levels can be higher in winter months, when we continue to make and eat cholesterol but there's less sunlight available to convert it.

It's interesting to note that, just as it blocks the ultraviolet rays that give us a suntan, sunblock also blocks the ultraviolet rays we need to make vitamin D. Australia recently embarked on an anti–skin cancer campaign it called "Slip-Slop-Slap." The campaign was especially effective at producing unintended results—Australian sun exposure went down, and Australian vitamin D deficiencies went up.

On the flip side, researchers have discovered that tanning

can actually help people who have vitamin D deficiencies. Crohn's disease is a disorder that includes significant inflammation of the small intestine. Among other things, the inflammation impairs the absorption of nutrients, including vitamin D. Most people who have Crohn's have a vitamin D deficiency. Some doctors are now prescribing UVB tanning beds three times a week for six months to get their patients' vitamin D back up to healthy levels!

Folic acid or folate, depending on its form, is just as important to human life. Folate gets its name from the Latin word for "leaf" because one of the best sources for folate is leafy greens like spinach and cabbage. Folate is an integral part of the cell growth system, helping the body to replicate DNA when cells divide. This, of course, is critical when humans are growing the fastest, especially during pregnancy. When a pregnant woman has too little folic acid, the fetus is at significantly higher risk for serious birth defects, including spina bifida, a deformation of the spinal cord that often causes paralysis. And as we said, ultraviolet light destroys folic acid in the body. In the mid-1990s an Argentinian pediatrician reported that three healthy women all gave birth to children who had neural tube defects after using indoor tanning beds during their pregnancies. Coincidence? Probably not.

Pregnancy isn't the only time folate is important, of course. A lack of folate is also directly linked to anemia, because folate helps to produce red blood cells.

THE SKIN, AS you've probably heard, is the largest organ of the human body. It's an organ in every sense of the word, responsible for important functions related to the immune system, the nervous system, the circulatory system, and metabolism. The skin protects the body's stores of folate, and it's in the skin that a crucial step in the manufacturing of vitamin D takes place.

As you might have guessed, the wide range of human skin color is related to the amount of sun a population has been exposed to over a long period. But darker skin isn't just an adaptation to protect against sunburn—it's an adaptation to protect against the loss of folic acid. The darker your skin, the less ultraviolet light you absorb.

Skin color is determined by the amount and type of melanin, a specialized pigment that absorbs light, produced by our bodies. Melanin comes in two forms—red or yellow pheomelanin, or brown or black eumelanin—and is manufactured by cells called *melanocytes*. Everybody on earth has around the same number of melanocytes—differences in skin color depend, first, on how productive these little melanin factories are and, second, on what type of melanin they make. The melanocytes of most Africans, for example, produce many times the amount of melanin that the melanocytes of Northern Europeans produce—and most of it is eumelanin, the brown or black version.

Melanin also determines hair and eye color. More melanin means darker hair and darker eyes. The milk white skin of an albino is caused by an enzyme deficiency that results in the production of little or no melanin. When you see the pink or red eyes that albinos usually have, you're actually seeing the blood vessels in the retina at the back of the eye, made visible by the lack of pigment in the iris.

As everybody knows, skin color changes, to some extent, in response to sun exposure. The trigger for that response is the pituitary gland. Under natural circumstances, almost as soon as you are exposed to the sun, your pituitary gland produces hormones that act as boosters for your melanocytes, and your melanocytes start producing melanin on overdrive. Unfortunately, it's very easy to disrupt that process. The pituitary gland gets its information from the optic nerve—when the optic nerve senses sunlight, it signals the pituitary gland to kick-start the melanocytes. Guess what happens when you're wearing sunglasses? Much less sunlight reaches the optic nerve, much less warning is sent to the pituitary gland, much less melanocyte-stimulating hormone is released, much less melanin is produced—and much more sunburn results. If you're reading this on the beach with your Ray-Bans on, do your skin a favor—take them off.

Tanning helps people cope with seasonal differences in sunlight in their ancestral climate; it's not enough protection for a Scandinavian at the equator. Someone like that—with very little natural ability to tan and regular, unprotected exposure to tropical sun—is vulnerable to severe burning, premature aging, and skin cancer, as well as folic acid deficiency and all its associated problems. And the consequences can be deadly. More than 60,000 Americans are diagnosed with melanoma—an especially aggressive type of skin cancer—every year. European Americans are ten to forty times as likely to get melanoma as African Americans.

AS HUMANITY WAS evolving, we probably had pretty light skin too, underneath a similar coat of coarse, dark hair. As we lost hair, the increased exposure of our skin to ultraviolet rays from the strong African sun threatened the stores of folate we need to produce healthy babies. And that created an evolutionary preference for darker skin, full of light-absorbing, folate-protecting melanin.

As some population groups moved northward, where sunlight was less frequent and less strong, that dark skin—"designed" to block UVB absorption—worked too well. Now, instead of protecting against the loss of folate, it was preventing the creation of vitamin D. And so the need to maximize the use of available sunlight in order to create sufficient vitamin D created a new evolutionary pressure, this time for lighter skin. Recent scientific sleuthing reported in the prestigious journal *Science* goes so far as to say that white-skinned people are actually black-skinned mutants who lost the ability to produce significant amounts of eumelanin.

Redheads, with their characteristic milky white skin and freckles, may be a further mutation along the same lines. In order to survive in places with infrequent and weak sunlight, such as in parts of the U.K., they may have evolved in a way that almost completely knocked out their body's ability to produce eumelanin, the brown or black pigment.

In 2000, an anthropologist named Nina G. Jablonski and a geographic computer specialist named George Chaplin combined their scientific disciplines (after already combining their lives in marriage) to chart the connection between skin color and sunlight. The results were as clear as the sky on a cloudless day—there was a near-constant correlation between skin color and sunlight exposure in populations that had remained in the same area for 500 years or more. They even produced an equation to express the relationship between a given population's skin color and its annual exposure to ultraviolet rays. (If you're feeling adventurous, the equation is $W = 70 - AUV/10$. W represents relative whiteness and

AUV represents annual ultraviolet exposure. The 70 is based on research that indicates that the whitest possible skin—the result of a
population that received zero exposure to UV—would reflect about
70 percent of the light directed at it.)

Interestingly, their research also proposes that we carry sufficient genes within our gene pool to ensure that, within 1,000 years
of a population's migration from one climate to another, its descendants would have skin color dark enough to protect folate
or light enough to maximize vitamin D production.

There is one notable exception to Jablonski and Chaplin's equation—and it's the exception that proves the rule. The Inuit—the
indigenous people of the subarctic—are dark-skinned, despite the
limited sunlight of their home. If you think something fishy's going on here, you're right. But the reason they don't need to evolve
the lighter skin necessary to ensure sufficient vitamin D production is refreshingly simple. Their diet is full of fatty fish—which
just happens to be one of the only foods in nature that is chock-full
of vitamin D. They eat vitamin D for breakfast, lunch, and dinner,
so they don't need to make it. If you ever had a grandmother from
the Old World try to force cod liver oil down your throat, she was
onto something for the same reason—since it's full of vitamin D,
cod liver oil was one of the best ways to prevent rickets, especially
before milk was routinely fortified with it.

IF YOU'RE WONDERING how people who have dark skin make
enough vitamin D despite the fact that their skin blocks all those
ultraviolet rays, you're asking the right questions. Remember, ultraviolet rays that penetrate the skin destroy folate—and ultraviolet rays that penetrate the skin are necessary to create vitamin D.

Dark skin evolved to protect folate, but it didn't evolve with a switch—you can't turn it off when you need to whip up a batch of vitamin D. So that would seem to create a new problem for people with dark skin—even if they lived in a sunny climate—because even though they received plenty of exposure to ultraviolet rays, the skin color that protected their supply of folate would *prevent* them from stocking up on vitamin D.

It's a good thing evolution's such a clever sort, because it took that into account—it kept room for a little guy called *apolipoprotein E (ApoE4)* in the gene pool of dark-skinned population groups. And guess what *ApoE4* does? It ensures that the amount of cholesterol flowing through your blood is cranked up. With more cholesterol available for conversion, dark-skinned people can maximize the use of whatever sunlight penetrates their skin.

Much farther to the north, without a similar adaptation, the light-skinned people of Europe would face a similar problem. There, instead of plenty of sunlight that was largely blocked by dark skin, they had to deal with too little sunlight to make enough vitamin D even with the benefit of their light skin. And sure enough, *ApoE4* is also common throughout Northern Europe. The farther north you go up the continent, the more you'll find it. As it does in Africans, the *ApoE4* gene keeps cholesterol levels cranked up, allowing its carriers to compensate for limited ultraviolet exposure by maximizing the cholesterol available for conversion to vitamin D.

Of course, in characteristic evolutionary fashion, *ApoE4* comes with a trade-off. The *ApoE4* gene and all the extra cholesterol that accompanies it put people at greater risk for heart disease and stroke. In Caucasians, it even carries a higher risk for development of Alzheimer's disease.

And as you've seen with iron loading and diabetes—one generation's evolutionary solution is another generation's evolutionary problem, especially when people no longer live in the environment that their bodies adapted to through evolution. (If you want a funny-sounding example of an environmental defense turned environmental hazard, you need look no further than your nose. ACHOO syndrome—its full name is autosomal dominant compelling helioopthalmic outburst syndrome—is the name of a "disorder" that causes uncontrolled sneezing when someone is exposed to bright light, usually sunlight, after being in the dark. Well, way back when our ancestors spent more time in caves, this reflex helped them to clear out any molds or microbes that might have lodged in their noses or upper respiratory tract. Today, of course, when someone is driving through a dark tunnel and emerges into the bright sun and gets a sneezing fit, ACHOO isn't helpful or funny at all—it can be downright dangerous.) But before we examine more instances of the effect a new environment has an old adaptations, let's take a look at another example of different population groups taking divergent evolutionary paths— this time, not just for environmental reasons, but for cultural reasons too.

IF YOU'RE OF Asian descent and have ever had an alcoholic beverage, there's a fifty-fifty chance your heart rate shot up, your temperature climbed, and your face turned bright red. If you're not Asian but you've ever been in a bar frequented by people with an Asian background, chances are you've seen this reaction. It's called Asian flush or, more formally, alcohol flush response. It happens to as many as half of all people of Asian descent, but it's uncommon in just about every other population group. So what's the story?

When you consume alcohol, your body detoxifies it and then extracts calories from it. It's a complex process that involves many different enzymes and multiple organs, although most of the process takes place in the liver. First, an enzyme called alcohol dehydrogenase converts the alcohol into another chemical called acetaldehyde; another enzyme—cleverly called acetaldehyde dehydrogenase—converts the acetaldehyde into acetate. And a third enzyme converts that into fat, carbon dioxide, and water. (The calories synthesized from alcohol are generally stored as fat—beer bellies really do come from beer.)

Many Asians have a genetic variation (labeled *ALDH2*2*) that causes them to produce a less powerful form of acetaldehyde dehydrogenase—one that isn't as effective in converting acetaledehyde, that first by-product of alcohol, into acetate. Acetaldehyde is thirty times as toxic as alcohol; even very small amounts can produce nasty reactions. And one of those reactions is the flushing response. That's not all it does, of course. After even one drink by people who have the *ALDH2*2* variation, the acetaldehyde buildup causes them to appear drunk; blood rushes to their face, chest, and neck; dizziness and extreme nausea set in—and the drinker is on the road to a nasty hangover. Of course, there's a side benefit to all this—people who have *ALDH2*2* are highly resistant to alcoholism. It's just too unpleasant for them to drink!

In fact, the resistance to alcoholism is so strong in people with *ALDH2*2* that doctors often prescribe alcoholics with a drug called disulfiram, which essentially mirrors the *ALDH2*2* effect. Disulfiram (Antabuse) interferes with the body's own supply of the acetaldehyde dehydrogenase enzyme, so anyone who drinks alcohol while taking it ends up with something that looks an awful lot like Asian flush and feels truly awful to boot.

So why is the *ALDH2*2* variation so common among Asians and virtually nonexistent among Europeans? It's all about clean water. As humans began to settle in cities and towns, they got their first taste of the sanitation and waste management problems that still plague cities today—but without even the possibility of modern plumbing. This made clean water a real challenge, and some theories suggest that different civilizations came up with different solutions. In Europe, they used fermentation—and the resulting alcohol killed microbes, even when, as was often the case, it was mixed with water. On the other side of the world, people purified their water by boiling it and making tea. As a result, there was evolutionary pressure in Europe to have the ability to drink, break down, and detoxify alcohol, while the pressure in Asia was a lot less.

Alcohol isn't the only beverage that requires some specific genetic mutation to enjoy, by the way. If you're reading this while sipping a latte or slurping an ice cream cone, you're a mutant. The great majority of the world's adults cannot eat or drink milk without experiencing a very unpleasant digestive reaction; once they no longer feed on breast milk, their bodies stop producing the enzyme that we need to digest lactose, the main sugar compound in milk. But if you can drink milk without the characteristic bloating, cramping, and diarrhea that signify lactose intolerance, you're a lucky mutant. You probably are descended from farmers who drank animal milk; somewhere along the line, a mutation sprang up that allowed people to keep producing the lactose-processing enzyme called lactase as adults, and that mutation spread throughout farming populations until it landed in your genome.

———

PEOPLE OF AFRICAN descent have darker skin and are much more likely to have a gene that causes them to produce greater amounts of cholesterol. People of Northern European descent have pale skin and are much more likely to have iron loading and a predisposition for Type 1 diabetes. People of Asian descent are much more likely to be unable to process alcohol efficiently. Are those *racial* differences?

It's not a question that can be easily answered. First of all, there's no real agreement as to what *race* means. On the genetic level, it's pretty clear that skin color isn't reliable. We've already discussed how the skin color of a transplanted population would change to match the level of ultraviolet exposure in its new environment. Recent genetic studies bear this out—in terms of common genetics, some dark-skinned North Africans are probably closer to light-skinned Southern Europeans than they are to other Africans with whom they share skin color.

On the other hand, many Jews seem to share a distinct genetic heritage despite the fact that they may be fair, blond, and blue-eyed or dark, black-haired, and brown-eyed. This has been borne out by recent research as well. Jews divide themselves into three groups to preserve certain religious traditions. The groups are based on which biblical tribe they are descended from—the Cohanim are members of the priestly tribe that traces its roots to Moses' brother Aaron, the original high priest. Levites are descendants of the tribe of Levi, the traditional princes of the temple. Today, descendants of the other twelve tribes are simply called Israelites.

A group of researchers recently compared the DNA of a large

group of Cohanim to the DNA of a large group of Israelites. The researchers were stunned to discover that—despite being spread across the world—the genetic markers of the Cohanim were so specific that they were all almost certainly descended from just a few male individuals. They came from Africa, from Asia, from Europe—and though their appearance ran the gamut from light-skinned and blue-eyed to dark-skinned and brown-eyed, most of them shared very similar Y chromosome markers. This controversial data even allowed the researchers to estimate when the originators of the Cohanim genes were alive. According to the researchers, that would have been 3,180 years ago, between the exodus from Egypt and the destruction of the First Temple in Jerusalem—or exactly when Aaron walked the earth.

NATURE GENETICS, A prominent journal, recently editorialized that "population clusters identified by genotype analysis seem to be more informative than those identified by skin color or self-declaration of race"—that makes a lot of sense. Instead of worrying about whether or not there are distinct "races," let's concentrate on what we do know and use that to advance medical science. What we do know is that distinct populations do share distinct genetic heritages, which are almost certainly the result of different evolutionary pressures our various ancestors experienced as they settled and resettled across the globe.

The current mainstream consensus is that modern humans evolved in Africa around 250,000 years ago. According to that theory, they migrated from Africa northward toward what is now the Middle East. Then some went right, populating India, the Asian coast, and ultimately, the Pacific Islands. Other groups headed left, settling across Central Europe. Still others continued north, spreading across Central Asia or venturing farther, by boat or by ice bridge, over the top of the world and then down into North and South America. All of that migration probably took place within the last 100,000 years. Of course, we don't know for sure yet. It's also possible that humans evolved in multiple places, and that different groups of prehumans and Neanderthals even interbred.

Whatever the truth is, it's clear that, as humanity evolved, different groups of humans encountered widely different circumstances—from infectious tropical diseases to sudden ice ages to pandemic plagues. The evolutionary pressure that accompanied all these challenges was probably intense enough to account for the differences we see between populations today. We've discussed a few examples, but the range is broad. Skull shape, for example, may have evolved as a mechanism to facilitate storage and release of heat depending on a population's climate.

Dense hair on the forearms and legs—the parts of the body usually exposed even with moderate dress—may have been a defense against malaria carried by mosquitoes. With the exception of Africa, where the heat was an evolutionary counterweight to thick body hair, the densest hair is generally found in the same places where malaria is most common—the eastern Mediterranean basin, southern Italy, Greece, and Turkey. In Africa, where the heat was an evolutionary argument against denser body hair, people are prone to sickle-cell anemia, which, as we'll discuss, offers some protection from malaria.

It's also important to remember that, in migratory terms, humanity has been on an express train for the last 500 years. The result, of course, is a blurring of genetic distinctions as people from different parts of the world meet and mate. Populations have always tended to combine genetic material (aka making babies) with nearby populations, but that genetic intermixing is taking place on a global scale today. In fact, genetic testing is revealing that the human population as a whole is already far more mixed than most people assume. Take Dr. Henry Louis Gates, for example, the distinguished scholar who is the chair of African and African American Studies at Harvard. Dr. Gates is black, but he and his family have long believed that they had at least one distant ancestor who wasn't black. Most likely a former slave owner who was thought to have been involved with his great-great-grandmother. And then some genetic testing revealed that Dr. Gates had no relationship to the slave owner—but fully 50 percent of his genetic heritage was European. Half of his ancestors were white.

Finally, we have to keep in mind that, in the right circumstances, heavy evolutionary pressure can breed a trait into—or out of—a population's gene pool in just a generation or two.

When you combine the possibility of relatively fast changes in a given gene pool with the rapid migration of the last 500 years, you can understand that population subsets with distinct genetic traits can emerge pretty quickly. A controversial theory looks to a shameful period in our history to explain the high rate of high blood pressure among African Americans.

High blood pressure, or hypertension, is a particularly insidious disease—it's responsible for as much as 25 percent of end stage kidney failure, but it usually has no noticeable symptoms; that's why it's often called the "silent killer." It is almost twice as common among African Americans as it is in the rest of the American population. Doctors first noticed the elevated incidence of high blood pressure in African Americans in the 1930s and assumed that all blacks shared a propensity for it. They were wrong. Blacks living in Africa do not have the same rate of hypertension as people of African descent in America. What's the explanation?

You've probably heard that salt can raise your blood pressure. Research has demonstrated that this is especially true for African Americans; their blood pressure is very reactive to salt. Now, salt also got a bad rap for a while, especially when it was first linked to high blood pressure, but it's a critical component of your body chemistry. It regulates fluid balance and nerve cell function. You can't survive without it. But when people who are especially reactive to it eat a diet high in salt, it can contribute to high blood pressure.

When Africans were taken to America against their will by slave traders, they were transported under horrible conditions— they usually weren't fed or even given sufficient amounts of water. The death rate was very high. It's possible that those with a natural propensity to retain high levels of salt had a better chance to survive—the extra salt helped them to maintain enough water to avoid fatal dehydration. If that's true, you can see how the slave trade might have produced a very *unnatural* selection for an increased ability to retain salt in many African Americans. When you couple that ability with a modern diet high in salt, it results in increased rates of hypertension.

FROM A MEDICAL perspective, it's clear that specific diseases are more prevalent in specific population groups in a way that is significant and deserves continued, serious exploration. On a proportional basis, African Americans have almost twice as many fatal heart attacks as European and South Asian Americans; their rate of cancer is 10 percent higher. European Americans are more likely to die of cancer and heart disease than Latino, Asian, or Native Americans. American Latinos are more likely to die of diabetes, liver disease, and infectious disease than non-Latinos. And Native Americans have higher rates of tuberculosis, pneumonia, and influenza. It seems like new examples crop up every month in the scientific literature. The most recent study discovered that African Americans who smoke a pack of cigarettes a day are far more likely to develop lung cancer than whites with the exact same habit.

Now, these statistics don't necessarily tell the whole story. For starters, they don't always control for other differences in these groups that have nothing to do with genetics and evolution. Differences in diet and nutrition, environment, personal habits, and access to health care will all have an effect on these studies. But that doesn't mean we should ignore the large trends we see among different population groups—to the contrary, the more we understand how our evolution has shaped our genetic makeup, the more we can understand how to live a healthy life today. Let's look at a few examples.

We've discussed two parallel adaptations to manage the sun's dueling effects on body chemistry—the evolution of dark skin to protect our stores of folate and the evolution of a genetic trigger for increased cholesterol to maximize production of vitamin D. Both of those adaptations are common in people of African descent and are effective—in the bright, strong sun of equatorial Africa.

But what happens when people with those adaptations move to New England, where the sun is much less plentiful and far less strong? Without enough sunlight to penetrate their dark skin and convert the additional cholesterol, they're doubly vulnerable—not enough vitamin D and too much cholesterol.

Sure enough, rickets—the disease caused by a vitamin D deficiency that causes poor bone growth in children—was very common in African American populations until we started routinely fortifying milk with vitamin D in the last century. And there appear to be connections among sunlight, vitamin D, and prostate cancer in African Americans as well. There is growing evidence that vitamin D inhibits the growth of cancerous cells in the prostate and in other areas, including the colon, too. Epidemiologists, who specialize in unlocking the mystery of where, why, and in whom disease occurs, have found that the risk of prostate cancer for black men in America climbs from south to north. When it comes to prostate cancer in black men, the risk is considerably lower in sunny Florida. But as you move north, the rate of prostate cancer in black men climbs until it peaks in the often cloud-covered heights of the Northeast. There is a growing belief among some researchers that a lack of vitamin D may also be one of the reasons we get sick more often in the winter than in the summer months.

The combination of excess cholesterol and lack of exposure to sufficient sunlight may well be part of the reason that African Americans have such a high rate of heart disease. The *ApoE4* gene keeps the blood full of cholesterol even though there's not enough sunlight in a northern climate to convert it to vitamin D. As cholesterol builds up, it attaches to the walls of your arteries—eventually, it can build up so much that it results in a blockage that causes a heart attack or a stroke.

The pharmaceutical industry has begun to take the genetic differences of populations into account. This study of how genetic variation can affect pharmaceutical treatment is called pharmacogenetics, and it's already producing results. There's a general consensus that some of the usual therapies for hypertension, for example, don't work as well for African Americans. The U.S. Food and Drug Administration (FDA) recently approved a controversial drug called BiDil for "self-identified" black patients who have heart failure.

New research has demonstrated that it's not just the presence of a specific genetic variation that can affect our body chemistry (and thus, the way we respond to a given drug)—it's how many times that gene occurs in our genome. In other words, it's quantity *and* quality.

For example, a gene called *CYP2D6* affects the way people metabolize more than 25 percent of all pharmaceuticals—including very common drugs like decongestants and antidepressants. People who have very few copies of this gene are called "slow metabolizers." It's thought that up to 10 percent of Caucasians fall into this category, but only 1 percent of Asians fit the bill. If you've ever taken a standard dose of Sudafed and felt a tingling sensation and a rapid heartbeat, you're probably a slow metabolizer, and you should talk to your doctor about cutting your dosage.

On the other end of the spectrum are ultrarapid metabolizers; these folks can have as many as thirteen copies of the *CYP2D6* gene! Of Ethiopians, 29 percent are metabolizers on hyperspeed, compared to less than 1 percent of Caucasians. The more we learn about the way genetic makeup affects an individual's response to a given drug, the more it's clear that "personalized medicine," where dosing and drugs are tailored to fit your genome, has the potential to provide significant health benefits.

Scientists suspect that the presence and quantity of genes like *CYP2D6* in different populations are related to the relative toxicity of a specific population's environment. Fast metabolizers can "clear"—detoxify—harmful substances more successfully. So the more toxins—from food, insects, whatever—in a particular environment, the more evolution favored multiple copies of toxin-clearing genes. Sometimes that fast metabolizing can be a problem too: some fast metabolizers actually convert certain drugs—like codeine—into much more potent forms. There was a recent report of a patient who became ill because she converted the codeine in her prescription cough syrup into morphine much faster than anyone expected. Sure enough, she was a *CYP2D6* fast metabolizer.

Another gene, this one called *CCR5-Δ32*, appears to prevent human immunodeficiency virus (HIV) from entering cells. One copy of this gene significantly hampers the virus's ability to multiply, reducing the viral load in people who carry the gene and become infected. And two copies of the gene? Almost complete immunity from HIV. Tragically, *CCR5-Δ32* is almost completely absent in Africans, where AIDS is epidemic, but it occurs in some 5 to 10 percent of Caucasians. Some researchers have suggested that *CCR5-Δ32* was selected for in the same way hemochromatosis was—because it offered some type of protection against the bubonic plague—but, unlike hemochromatosis, no clear mechanism for this selection has been suggested.

One thing is clear—there is mounting evidence that where our ancestors came from, how they adapted to manage their environment, and where we live today all combine to have a significant impact on our health. That understanding ought to inform everything from research in the laboratory to medical care in the doctor's office to life in our homes. Today, the most widely prescribed therapy for high cholesterol is a class of drugs called *statins*. Although they are considered generally "safe" drugs, over time, statins can cause serious side effects, including liver damage. If you knew that you might be able to reduce your excess cholesterol by getting enough sunlight to convert it to vitamin D, wouldn't you rather hit the tanning salon before starting a lifetime of Lipitor?

That's food for thought.

CHAPTER IV

HEY, BUD, CAN YOU DO ME A FAVA?

A distinguished-looking man, debonair to his core in a way that the bright orange prison coveralls cannot obscure, stands in his jail cell looking out at an attractive brunette who has presumed—presumed!—to question him. She's testing him—and he's having none of it. "A census taker once tried to test me. I ate his liver with some fava beans and a nice Chianti," says Hannibal Lecter.

If the doctor they called the cannibal had been an epidemiologist instead of a psychiatrist, he might have killed his victim with those fava beans—not just served his liver with them.

Before we started calling them fava beans, after the Italian word for them, we called them broad beans—and the range of legend that surrounds them is certainly broad. The Greek scholar Pythagoras supposedly warned a flock of future philosophers, "Avoid fava beans." Of course, since fava beans were used as ballots at the time—white for yes and black for no—he may have just been giving his students advice that all good philosophers should still ponder today—"Avoid politics."

In fact, the legends surrounding Pythagoras's warning are almost as varied as the legends around the bean itself. A different theory holds that Pythagoras's concern was something much less grave than possible poison and much less theoretical than possible politics—according to Diogenes, Pythagoras was just worried his students would eat too many beans and, well, pass too much gas. Two thousand years ago Diogenes supposedly said:

> One should abstain from fava beans, since they are full of wind and take part in the soul, and if one abstains from them one's stomach will be less noisy and one's dreams will be less oppressive and calmer.

A cult called the Orphics believed that the fava plant contained the souls of the dead: according to them, "Eating fava beans and gnawing on the heads of one's parents are one in the same." Aristotle alone had five different theories about Pythagoras's broad beans, saying that Pythagoras warned against them

> either because they have the shape of testicles, or because they resemble the gates of hell, for they alone have no hinges, or again because they spoil, or because they resemble the nature of the universe, or because of oligarchy, for they are used for drawing lots.

It's no wonder all those ancient Greeks were philosophers—they clearly had a lot of time on their hands. But they weren't the only people to notice the mysterious reaction many people have to fava beans. In the twentieth century, a schoolteacher in Sardinia, an island off the coast of Italy, is said to have noticed a seasonal lethargy that settled on her students every spring and lasted for weeks. Supposedly recalling Pythagoras's warning, she connected her students' nodding heads to flowering fava plants. Superstitions against eating uncooked fava beans were common throughout the Middle East. In Italy, fava beans are traditionally planted on All Souls' Day, and cakes shaped like a fava bean pod are called *fave dei morti*—"beans of the dead."

As you've probably come to suspect, where there's folklore smoke, there's medical fire—in the case of the fava bean, a whole lot of it.

Favism, as modern medical science has so aptly labeled it, is an inherited enzyme deficiency carried by 400 million people. It's the most common enzyme deficiency in the world. In extreme cases, people who have favism and eat fava beans (or take certain drugs) experience rapid, severe anemia that can often lead to death.

SCIENTISTS FIRST CAUGHT wind of the truth behind some people's deadly reaction to fava beans during the Korean War. Because malaria was common in parts of Korea, American soldiers who served there were prescribed antimalarial drugs, including one called primaquine. Doctors soon discovered that about 10 percent of African American soldiers developed anemia while taking primaquine, and some soldiers, especially those of Mediterranean descent, experienced an even more severe side effect called hemolytic anemia—their red blood cells were literally bursting.

In 1956, three years after the ceasefire that ended the Korean War, medical researchers isolated the cause of the soldiers' reaction to the antimalarial drugs—they lacked sufficient amounts of an enzyme called glucose-6-phosphate dehydrogenase, or G6PD for short. G6PD is thought to be present in every cell in the body. It's especially important in red blood cells, where it protects cellular integrity, mopping up chemical elements that would otherwise destroy the cell.

You've probably heard about free radicals in the news and may have a general sense that they're not so good for you. The easiest way to understand free radicals is to remember that Mother Nature likes pairs—she's something of a chemical matchmaker. Free radicals are essentially molecules or atoms with unpaired electrons—and unpaired electrons look to pair up. Unfortunately, as far as your body is concerned, those electrons look for love in all the wrong places. As the unpaired electrons seek to pair with electrons in other molecules, they cause chemical reactions. Those reactions can disrupt cellular chemistry and lead to the cell's early death. That's one of the reasons free radicals are thought to be a major cause of aging.

G6PD is like a bouncer in the red blood cell bar: when it's on the job, it throws out the free radicals so they can't start trouble. But when you don't have enough G6PD, any chemical that produces free radicals can wreak havoc on your red blood cells. That's what happened with the soldiers who experienced adverse reactions to primaquine—one of the ways primaquine is thought to stop the spread of malaria is by stressing your red blood cells and making them a generally unpleasant place for malaria-causing parasites. But if you don't have enough G6PD to maintain cellular integrity, when the primaquine puts stress on your red blood cells, some of the cells can't take it—the free radicals cause the cell membranes to burst, destroying them. And that loss of red blood cells spells anemia—specifically, hemolytic anemia, which is anemia that is caused by the early breakdown of red blood cells. The person undergoing the hemolytic crisis will experience severe weakness and fatigue; there may be signs of jaundice. Untreated, hemolytic anemia can lead to kidney failure, heart failure, and death.

THOSE ANCIENT GREEKS were onto something—for some people, fava beans are killers. They contain two sugar-related compounds called vicine and convicine. Vicine and convicine produce free radicals, especially hydrogen peroxide. When people who have favism eat fava beans, they undergo a reaction similar to the one that occurs after taking primaquine. If the hydrogen peroxide isn't cleared out with the help of G6PD, it starts to attack your red blood cells, ultimately breaking them down. When that happens, the rest of the cell leaks out, resulting in hemolytic anemia, with potentially deadly effect.

The gene that is responsible for G6PD protein production—or deficiency—goes by the same name, G6PD. This gene is carried on the X chromosome. As you probably remember from science class, the X chromosome is one of the two sex chromosomes; the other is Y. Humans with two X chromosomes—XX—are female; humans with an X and a Y chromosome—XY—are male. Because the gene for G6PD deficiency is carried on the X chromosome, the condition is much more common in men. When a man has the mutation on his one X chromosome, all his cells take direction from that mutation. For a woman to have serious G6PD deficiency, she has to have the mutation on both X chromosomes. If she has it on only one chromosome, some of her red blood cells will have a normal gene and some won't, and she should produce sufficient G6PD to avoid favism.

There are two normal versions of the *G6PD* gene, one called Gd^B and the other Gd^{A+}. There are more than 100 possible mutations of this gene, but they fit into two major categories, one that arose in Africa, called Gd^{A-}, and one that arose around the Mediterranean, called Gd^{Med}. These mutations cause serious problems only when free radicals start overwhelming your red blood cells and there isn't enough G6PD to clean them up. Problems can be triggered in people with favism by some infections and some medications—like primaquine—that release free radicals into the bloodstream. But as we've discussed, the most common trigger is eating fava beans—which is why it's called favism, of course.

Humans have been cultivating fava beans for thousands of years. The oldest seeds found so far were discovered in an archaeological dig near Nazareth. They're thought to be around 8,500 years old, having been dated back to 6500 B.C. From Nazareth, in what is now the northern part of Israel, fava beans are thought to have spread throughout the Middle East and then north, around the eastern Mediterranean, into Turkey, across the Greek plains, and on into southern Italy, Sicily, and Sardinia.

If you marked up a map to show where favism is most common, and then overlaid that with the areas where fava bean cultivation is most common, guess what? At this point, you may not be all that surprised at what I'm about to tell you—favism genes and fava bean farms? Same places, same people. Favism is most common—

and most deadly—in North Africa and Southern Europe, all around the Mediterranean. Which happen to be exactly the places where fava beans are historically cultivated and consumed.

Here we go again—somehow millions of humans have evolved with a genetic mutation that is only likely to cause problems when they eat something that is *most common to the diet in their part of the world?*

Well, if we've figured out anything so far, it's that evolution doesn't favor genetic traits that will make us sick unless those traits are more likely to help us before they hurt us. And a trait that is shared by more than 400 million people is definitely an evolutionary favorite. So there has to be some benefit to G6PD deficiency, right?

Right.

BEFORE WE DIG further into the connection between favism and fava beans, let's take a look at the broader connection between evolution in the animal kingdom and evolution in the plant kingdom. We'll start with breakfast. You see that strawberry in your cereal? The vine it came from wants you to eat it!

Plants that produce edible fruit evolved that way for their own benefit. Animals pick fruit and eat it. The fruit contains seeds. Animals walk or lope or swing or fly away and eventually they *deposit* those seeds somewhere else, giving the plant a chance to spread and reproduce. The apple doesn't fall far from the tree— unless an animal eats it and takes it for a ride. It's a gastronomic hitchhike, and it works well for everybody. In fact, that's why ripe fruit is easy to pick and often falls off while unripe fruit is harder to harvest—the plant doesn't want you to take off with the fruit until the seeds within it have finished developing. No free lunch in Mother Nature's outdoor cafe.

On the other hand, as much as plants want animals to eat their fruit, they don't want animals to get much closer than that—when creatures start to nibble on their leaves or gnaw at their roots, things can get tricky. So plants have to be able to defend themselves. Just because they're generally immobile doesn't mean they're pushovers.

Thorns are plants' most obvious defense mechanism, but they're by no means the only one, or the most powerful—these guys have a whole arsenal. Plants by far are the biggest manufacturers of chemical weapons on earth. Everybody knows about the beneficial effects we receive because of basic plant chemistry. They convert sunlight and water into sugar by using carbon dioxide they absorb from the atmosphere, in turn producing oxygen, which we get to breathe. But that's just the starting point. Plant chemistry has the power to make a significant impact on its environment, influencing everything from the weather to the number of local predators.

Clover, sweet potato, and soy all belong to a group of plants that contain a class of chemicals called phytoestrogens. Sounds familiar, right? It should. Phytoestrogens mimic the effect of animal sex hormones such as estrogen. When animals eat too much of a plant that contains phytoestrogens, the overload of estrogenlike compounds wreaks havoc on their reproductive capability.

There was a sheep-breeding crisis in Western Australia during the 1940s. Otherwise healthy sheep weren't getting pregnant or were losing their young before giving birth. Everyone was stumped until some bright agricultural specialists discovered the little culprit—European clover. This type of clover produces a potent phytoestrogen called formononetin as a natural defense against grazing predators. And, yes, if you're a plant, a sheep is a predator! Accustomed to the humidity of Europe, the imported clover plants were struggling to cope with the drier Australian climate. When clover has a bad year—not enough rain or sunshine, or too much rain or sunshine—it protects itself by limiting the size of the next generation of predators. It increases production of formononetin and prevents the birth of baby grazers by sterilizing their would-be parents.

The next time you're looking for some convenient birth control, you don't have to snack on a field of clover, of course. But if you take many forms of the famous "Pill," you're not doing something all that different. The gifted chemist Carl Djerassi based his development of the Pill on just this kind of botanical birth control. He wasn't using clover, though; he was using sweet potatoes—the Mexican yam to be exact. He started with disogenin, a phytoestrogen produced by the yam, and from that base, he synthesized the first marketable contraceptive pill in 1951.

Yams aren't the only source of phytoestrogens in the human diet. Soy is rich in a phytoestrogen called genistein. It's worth noting that today many processed foods, including commercial baby formulas, use soy because it's an inexpensive source of nutrition. There's a growing concern among a small number of scientists that we don't have a handle on the potential long-term effects of what seems to be an ever-greater level of phytoestrogens and soy in our diet.

PLANTS ARE GOOD at birth control—but they're great at poison. Most of the toxins they produce aren't directed at humans, of course; they don't really have to worry about us too much. The real problem that plants have is all those committed vegetarians grazing and buzzing and flying around that rely solely on them for food. But that doesn't mean we don't have to be careful, because plant toxins can cause lots of problems for us too. And chances are, you've probably eaten your fair share in the last week.

Ever have tapioca pudding? Tapioca is made from the cassava plant. Cassava is a large, thick-skinned tuber that looks kind of like a long white sweet potato with a coconut's skin. It's a major part of the diet in many tropical countries. Yet it contains a precursor to deadly cyanide. Of course, when it's cooked and processed correctly, it can be harmless—so don't go biting down on the next raw cassava plant you see. Not surprisingly, cassava is especially high in cyanide compounds during drought—exactly when it needs additional protection against predators to make it through the growing season.

Consider another example, the Indian vetch, which is cultivated in Asia and Africa. Its chemical weapon of choice is a powerful neurotoxin that can cause paralysis. The neurotoxin is so powerful that the vetch can often survive when all other crops die out because of drought or infestation. For that reason, poor farmers in some parts of the world cultivate it as an insurance crop—insurance against starvation in the event of a famine. And sure enough, the incidence of disease related to this organic poison climbs after a famine in those areas where the vetch is grown. Not surprisingly, some people choose to risk the vetch's poison rather than starve to death.

The nightshades are a large group of plants, some edible, some poisonous. All nightshade contains a large portion of alkaloids, chemical compounds that can be toxic to insects and other herbi-

vores and affect humans in ways ranging from helpful to halluci-
nogenic. Some people speculate that "witches" included some types
of nightshade in their "magic" ointments and potions—and then
hallucinated that they were flying!

One of the most common members of the nightshade family,
which includes potatoes, tomatoes, and eggplant, is the jimsonweed, which got its name from Jamestown, Virginia. About a hundred years before the Revolutionary War there was a short-lived
revolt called Bacon's Rebellion. It was defeated pretty quickly, but
not without some hiccups along the way. When British soldiers
were sent to Jamestown to put down the rebellion, they were se-
cretly (or accidentally) drugged with jimsonweed in their salad. In
1705 Robert Beverley described the result in *The History and Present State of Virginia:*

> Some of them ate plentifully of it, the effect of which was a
> very pleasant comedy, for they turned natural fools upon it for
> several days: one would blow up a feather in the air; another
> would dart straws at it with much fury; and another, stark na-
> ked, was sitting up in a corner like a monkey, grinning and
> making mows at them; a fourth would fondly kiss and paw his
> companions, and sneer in their faces with a countenance more
> antic than any in a Dutch droll. . . . A thousand such simple
> tricks they played, and after 11 days returned themselves again,
> not remembering anything that had passed.

Jimsonweed is a tall green weed with big leaves that is common
throughout America. People eat it accidentally every year, usually
because it's mixed in with other plants in their garden.

Plant chemicals can paralyze, sterilize, or make us crazy. They

can also affect us in much milder ways, like interfering with diges-
tion or burning our lips. Wheat, beans, and potatoes all have amy-
lase inhibitors, a class of chemicals that interfere with the absorption
of carbohydrates. Protease inhibitors, found in chickpeas and some
grains, interfere with protein absorption. Many of these botanical
defense systems can be disabled by cooking or soaking. The Old
World tradition of soaking beans and legumes overnight does exactly that—it neutralizes most of the chemicals that mess with our
metabolism.

If you've ever bitten down on a raw habanero pepper, you probably felt like you were being poisoned. And you were. That burning
sensation is caused by a chemical called capsaicin. Mammals are
sensitive to it because it tickles the nerve fibers that sense pain and
heat, but birds aren't—and that goes to show just how clever old
Mother Nature can be when she's doing the evolution dance. Mice
and other rodents that would otherwise be drawn to the fruits of
chili plants avoid them because they can't take the heat. That's good
for the chili, because the digestive systems of mammals destroy its
small seeds, a process that pretty much takes the point out of the
gastronomic hitchhike. Birds, on the other hand, don't destroy chili
seeds when they eat chili peppers—and they aren't affected by capsaicin. So mammals leave the peppers for the birds, and the birds
take the seeds to the air, spreading them along the way.

Capsaicin is a sticky poison—it adheres to mucous membranes,
which is why your eyes burned if you ever rubbed them after handling peppers. It's also why the heat from a hot pepper sticks
around so long—and why water does nothing to cool the burn. Its
stickiness acts to prevent capsaicin from easily dissolving in water.
You're much better off drinking milk (but this is one time to pass

on the skim!) or eating something else with fat in it—since fat is hydrophobic, it helps to peel the capsaicin away from your mucous membranes and cool you down.

Capsaicin doesn't just cause a burning sensation—it can actually cause selective degeneration of some types of neurons. In large quantities, hot peppers can be very harmful. Scientists are still debating the connection, but people in places like Sri Lanka where hot peppers are almost a staple, as well as other ethnic groups who eat lots of hot peppers, tend to have much higher rates of stomach cancer.

From an evolutionary perspective, it's not surprising that plants have evolved mechanisms to ensure that their predators think twice before making them their next meal. What's more surprising is why we continue to cultivate and consume thousands of plants that are toxic to us. The average human eats somewhere between 5,000 and 10,000 *natural* toxins every year. Researchers estimate that nearly 20 percent of cancer-related deaths are caused by natural ingredients in our diet. So if many plants we cultivate are toxic, why didn't we evolve mechanisms to manage those toxins or just stop cultivating them?

Well, we have.

Sort of.

HOW MANY TIMES have you had a craving for something sweet? Or something salty? How about something bitter? Can't you just see yourself saying, "Man, all I really want is something really bitter for dinner." Doesn't happen, right?

There are four basic tastes in Western tradition—sweet, salty,

sour, and bitter. (There's a fifth in other parts of the world that is
gaining traction in the West, both culturally and scientifically—it's
called *umami,* and it's the savory flavor you find in aged and fer-
mented foods, like miso, parmesan cheese, or aged steaks.) Most
tastes are pleasing, and the evolutionary reason for them is simple—they attract us to foods that contain nutrients, as well as the
salt and sugar, that we need.

Bitterness is different—bitterness turns us off. Which, as it
turns out, is probably the point. A study published in 2005 by researchers working collaboratively at University College London,
Duke University Medical Center, and the German Institute of
Human Nutrition concluded that we evolved the ability to taste
bitterness in order to detect toxins in plants and avoid eating them.
(Which is why the plants produce the toxins in the first place and
has led to the term many plant biologists use to describe them—
antifeedants.) By reconstructing the genetic history of one of the
genes responsible for the growth of bitter taste receptors in our
tongues, scientists have traced the evolution of this ability to Africa, sometime between 100,000 and 1,000,000 years ago. Not all
humans have the ability to taste bitterness—and not all are as sensitive to it as others—but given how widespread the ability is across
the globe, it's pretty clear that tasting bitterness gave humans a
significant survival advantage.

About one-quarter of humanity is even more highly attuned to
taste. They're called supertasters—because they are. Chemists discovered them almost by accident while studying reactions to a
chemical called propylthiouracil. Some people can't taste it at all.
Some people find it to be mildly bitter. And some people—supertasters—find even the smallest taste to be repulsive. Supertasters
find more bitterness in grapefruit, coffee, and tea. They may be as

much as twice as sensitive to sweetness and are much more likely to feel the fire at a hint of chili.

Interestingly, the same collaborative paper that linked bitterness to the detection of plant poisons noted that it may not be such an advantage today. Not every version of the compounds that taste bitter is poisonous; in fact, as I mentioned in the description of nightshade, some of these compounds are beneficial. The scopolamine in jimsonweed that causes temporary madness is a bitter-tasting alkaloid—but so are some of the compounds in broccoli that have anticancer properties. So today, especially in developed countries where the need for a natural alarm bell against plant toxins has pretty much faded away, it may be a disadvantage to have a strong reaction to bitterness. Now, instead of steering you away from poisons, it's steering you away from food that's good for you.

WITH A QUARTER-MILLION plants to choose from and a keen sense of taste, why haven't we cultivated plants that aren't poisonous and bred the toxins out of plants that are? Well, we've tried—but like everything else in the evolutionary kingdom, it's complicated. And there are consequences.

Remember, plants' chemical weapons aren't aimed at us for the most part; they're directed more at insects, bacteria, fungi, and, in some cases, mammals that are dedicated herbivores. So if we impose unilateral disarmament on a plant, it's like giving the keys to the candy store to a busload of schoolkids—pretty soon there's nothing left for anyone else to eat. The plant's predators just finish it off.

Of course, sometimes plant breeders have gone the other way

and bred in too much natural resistance, turning an otherwise edible food into an almost deadly poison. All potatoes contain solanine, especially those that are a little green in color. Solanine is also what protects potatoes from potato late blight (imagine a deadly case of athlete's foot and you'll get the idea of what blight means to a potato). Solanine is a fat-soluble toxin that can cause hallucinations, paralysis, jaundice, and death. Too many solanine-rich french fries and you're french fried. Sometimes, of course, blight can overwhelm the protection solanine provides. The fungus was responsible for the devastating Irish potato famine in the mid-nineteenth century that led to mass starvation, death, and emigration from Ireland.

In England during the 1960s, plant breeders worked to develop a blight-resistant potato, in order to increase the efficiency of potato crops. They called their special spud the Lenape. The first person who ate a Lenape potato didn't feel very special, though—it contained so much solanine, it was nearly deadly. You won't be surprised to hear that they pulled those Lenapes from the market like—hot potatoes.

Celery is a similar case that sheds light on the sometimes double-edged nature of organic agriculture. Celery defends itself by producing psoralen, a toxin that can damage DNA and tissue and also causes extreme sensitivity to sunlight in humans. The funny thing about psoralen is that it becomes active only when it's exposed to sunlight. Some insects avoid this poison by keeping their victim in the dark—they roll themselves up in a leaf, protected from the sun, and then spend the day chewing their way out.

Garden-variety celery doesn't pose a problem to most people, unless you visit the tanning salon after a bowl of celery soup. Psoralen generally poses more of a problem for those who handle large amounts of celery over a long period of time—many celery pickers have developed skin problems, for example.

Now, the thing about celery is that it's especially good at kicking psoralen production into high gear when it feels under attack. Bruised stalks of celery can have 100 times the amount of psoralen of untouched stalks. Farmers who use synthetic pesticides, while creating a whole host of other problems, are essentially protecting plants from attack. Organic farmers don't use synthetic pesticides. So that means organic celery farmers are leaving their growing stalks vulnerable to attack by insects and fungi—and when those stalks are inevitably munched on, they respond by producing massive amounts of psoralen. By keeping poison *off* the plant, the organic celery farmer is all but guaranteeing a biological process that will end with lots of poison *in* the plant.

Life: it's such a compromise.

NOW THAT WE have a better understanding of the relationship that plant evolution has on humans, let's take another look at the connection between fava beans and favism.

What do we know so far? We know that eating fava beans releases free radicals into the bloodstream. We know that people who have favism, with a deficiency in the G6PD enzyme, lack the ability to mop up those free radicals, which causes their red blood cells to break down and results in anemia. We know that a map of fava bean cultivators and a map of likely favism carriers would highlight the same portions of the globe. And we know that any genetic mutation as common as favism—more than 400 million people—must have given its carriers some advantage over something even more deadly.

So what's a threat to human survival that is common in Africa and around the Mediterranean and has a connection to red blood cells? Four out of five dentists may recommend Trident—but ten out of ten infectious-disease experts will give you the same answer if you ask them to solve that riddle: the answer is malaria.

Malaria is an infectious disease that infects as many as 500 million people every year, killing more than 1 million of them. More than half of the world's population live in areas where malaria is common. If you're infected, you can experience a terrible cycle of fevers and chills, along with joint pain, vomiting, and anemia. Ultimately, it can lead to coma and death, especially in children and pregnant women.

For centuries, starting with Hippocrates' treatise *On Airs, Waters, and Places,* doctors believed many diseases were caused by unhealthy vapors emanating from still water—lakes, marshes, and swamps. They called these vapors or mists *miasma.* Malaria, which is Old Italian for "bad air," was one of many diseases they thought miasma caused. The association with hot, wet swamps turned out to be correct—but because of the mosquitoes that thrive in those areas, not the vapors they emit. Malaria is actually caused by parasitic protozoa (microsopic organisms that share some traits with animals) that are deposited in the human bloodstream through the bite of female mosquitoes (males don't bite). There are a few different species that cause malaria, the most dangerous of which is *Plasmodium falciparum.*

The theory that miasma causes malaria was wrong, but it led to the development of at least one modern comfort many people would sweat to lose. According to James Burke, the author of the *Connections* series, a Florida doctor named John Gorrie thought he had malaria licked in 1850, with the help of a new invention. Dr. Gorrie correctly noticed that malaria was significantly more common in warmer climates. And even in cooler places, people seemed to get sick only in warmer months. So he figured if he could find a way to eliminate all the warm "bad air," he could protect people from malaria.

Dr. Gorrie's malaria-fighting contraption pumped cool air into the malaria hospital ward. Today, a version of his invention probably pumps cool air into your home—you call it an air conditioner. And while the air conditioner didn't improve the prognosis of any of Dr. Gorrie's malaria-infected patients, it has had an impact on the disease. Air-conditioning allows people who live in malarial parts of the world to stay inside with their doors closed and windows shut, which helps to protect them from infected mosquitoes.

There are still hundreds of millions of malarial infections every year—and while it's one of the ten highest causes of death in the world, not everybody who gets infected dies. Even more to the point, perhaps, not everyone who gets bit by malaria-carrying mosquitoes gets infected. So what's helping the malaria survivors?

J. B. S. HALDANE was one of the first people to understand the idea that different environments impose different evolutionary pressures, producing distinct genetic traits that in certain populations cause disease. More than fifty years ago, he suggested that certain groups—specifically people with a genetic tendency for sickle-cell anemia or thalassemia, another inherited blood disorder—had better natural resistance to malaria.

Today many researchers believe that a genetic trait far more prevalent than sickle-cell anemia or thalassemia may also provide protection against malaria—G6PD deficiency. In two large case-controlled studies, researchers found that children with the African variant of the G6PD mutation had twice the resistance to *P. falciparum,* the most severe type of malaria, that children without the mutation had. Laboratory experiments confirmed this— given a choice between "normal" red blood cells or G6PD-deficient red blood cells, the malarial-causing parasites preferred the normal cells time after time.

Why? *P. falciparum* is actually a delicate little creature. It only really thrives in nice clean red blood cells. The red blood cells of someone with G6PD are not just less hospitable to malaria, they are also taken out of circulation sooner than those of people without the mutation, and that disrupts the parasite's life cycle. This explains why populations exposed to malaria would select for favism. What it doesn't explain is why those populations would also cultivate fava beans. What's the point in living through a mosquito's breakfast if your own lunch can kill you?

The answer is probably straightforward—redundancy. Malaria is so widespread and so deadly that vulnerable populations needed every possible defense in order to survive and reproduce. By releasing free radicals and raising the level of oxidants, fava bean consumption makes the blood cells of non–G6PD deficient people a less hospitable place for malarial parasites. With all the free radicals, some red blood cells tend to break down. And when someone with a mild or *partial* deficiency in G6PD eats fava beans, the parasite is in deep trouble.

As far as partial deficiency is concerned, remember that the genetic mutation that causes favism is only passed on the X chromosome, and remember that females have two X chromosomes. That means that (in populations where the mutation is common) many women have a red blood supply that is partially normal and partially G6PD deficient. That gives them additional protection against malaria, but doesn't make them vulnerable to an extreme reaction to fava beans. And considering that pregnant women are very vulnerable to malaria, it's a good thing that many women can have their favism and eat it too.

HUMANS HAVE BEEN relying on herbal remedies since, well, probably before there were humans. Archaeologists have found evidence suggesting that Neanderthals may have used plants for healing 60,000 years ago. The ancient Greeks used opium milk, which is the fluid that oozes out of the opium poppy when it's slashed, as a painkiller—today we derive morphine, one of the most powerful painkillers available, from the same place.

The first really effective antimalarial medicine came from the bark of the cinchona tree. George Cleghorn, a Scottish army surgeon, is one of those credited with discovering the antimalarial properties of cinchona bark early in the nineteenth century, but it still took another century before French chemists isolated the specific beneficial compound—quinine—and made a medicinal tonic from it. The tonic tasted awful, though, so legend has it that British soldiers mixed their gin rations with their tonic treatments and presto, a classic was born. Tonic water still contains quinine today, but unfortunately, if you're going to travel somewhere that malaria is prevalent, you still need a prescription for your antima-

larial drug; just about every strain of malaria has become somewhat resistant to quinine. Good thing we have those helpful fava
beans.

Eat your vegetables. Your vegetables can kill you.

Mother Nature is sending mixed messages again. The truth—as
you've no doubt gathered—is complicated. Many plant toxins can
be good for us. The trick is understanding how they work, how we
work, and how it all works together.

Those phytoestrogens that can cause sterility? It looks like genistein, the phtyoestrogen in soy, might help to stop or slow the
growth of prostate cancer cells. Some researchers think the same
compound may ease the effects of menopause, which could explain
why Asian women report far fewer problems with mid-life
changes.

Capsaicin, the hot in hot peppers, stimulates the release of en-
dorphins, which induce feelings of pleasure and reduce feelings of
stress. Capsaicin also increases your metabolic rate—some think
by as much as 25 percent. Even more, there is a growing body of
evidence that capsaicin may be helpful in alleviating pain caused
by everything from arthritis and shingles to postoperative discom-
fort.

The list goes on. The psoralen in celery can cause skin damage—but it also is a real help for people with psoriasis. Allicin,
which comes from garlic, prevents platelets in your blood from
sticking together and becoming clots, which makes it a potentially
powerful weapon against heart disease. The aspirin a day that keeps
the doctor away? It started out as a chemical in the bark of willow
trees to keep the insects away. Today it's a virtual drug of all
trades—a blood-thinning, fever-reducing pain reliever. Taxol? The

powerful anticancer drug is another tree bark derivative—in this case from the bark of the Pacific yew.

Around 60 percent—or more—of the world's population still relies directly on plants for medicine. Probably isn't such a bad idea for us to drop in every once in a while, take a look at what they're cooking, and wonder why.

CHAPTER V

OF MICROBES AND MEN

For thousands of years a parasitic worm called *Dracunculus medinensis*—which means "little dragon"—has plagued humans across Africa and Asia. It causes a terrible disease. The larvae of the worm, also known as Guinea worm, are eaten by water fleas that fill ponds and other sources of still water in remote tropical areas. When people drink the water, their digestive system destroys the fleas but not the larvae. Some of the larvae migrate from the small intestine into the body, where they grow and eventually mate with each other. About a year after infection, adult females—now two to three feet long, about the diameter of a piece of spaghetti, and full of new larvae themselves—make their way to the skin of the person carrying them. Once they get to the surface, these female Guinea worms begin to secrete acid, effectively burning themselves an exit tunnel. The first sign of infection is the appearance of a painful blister. Soon after the blister appears it ruptures painfully, and the worm starts to make its way out. The burning caused by the acid drives the human host to seek relief in cooling water. And as soon as the worm senses water it emits a milky fluid full of thousands of larvae to begin the process anew.

Worms can sometimes be removed surgically, but for millennia, the only effective treatment has been to wrap the worm around a stick and slowly, carefully pull it out. The process lasts for many painful weeks or months and can't be hurried along too quickly—if the worm breaks, the infected person can experience an even more painful and serious reaction, perhaps even death.

Guinea worm has afflicted humanity for centuries. It's been found in Egyptian mummies and even thought to be the "fiery serpent" that ravaged the Israelites during their forty years in the desert. Some scholars think the Rod of Asclepius—the snake wrapped around a staff that is a symbol of medicine—was originally a simple drawing that early doctors used to show they offered help to remove the worms by wrapping them around a stick.

Today, because we understand how the Guinea worm manipulates its victims to collaborate in the infection of others, the little dragon's fire is on the verge of being extinguished. Former president Jimmy Carter has led a two-decade effort to spread understanding about the parasite's method of reproduction to every corner of the world, ensuring that its victims avoid water when looking for relief and that its potential victims avoid water that could be infected. According to the Carter Center, the worldwide incidence of Guinea worm infections had dropped from 3.5 million in 1986 to just 10,674 in 2005. By understanding how the

Guinea worm has evolved in relationship to us, we have the chance to protect people from it.

IF YOU'VE COME this far on our journey across the evolutionary landscape, you've probably gathered a good sense of the interconnectedness of—well, just about everything. Our genetic makeup has been adapting in response to where we live and what the weather's like. The food we eat has evolved to cope with the organisms that eat it, and we've evolved to cope with that. We've looked at the way we've evolved to resist or manage the threat posed by specific infectious diseases, like malaria. But what we haven't discussed—yet—is how all those infectious diseases are evolving right along with us. Make no mistake—they are, and for the exact same reason that we've been evolving for millions of years too. At the end of the day, every living thing—bacteria, protozoa, lions, tigers, bears, and your baby brother—shares two hardwired imperatives: Survive. Reproduce.

Now, in order to really understand the relationship between humans and the millions of microbes living beside us, you have to discard the notion that all bacteria are bad, all microbes are marauders, all viruses are villains, all—okay, you get the point. The truth is that we have been evolving in tandem with all of these microscopic organisms—often to our mutual benefit. The way our bodies work today is directly related to our interaction with infectious agents over millions of years. Everything from our senses to our appearance to our blood chemistry has been shaped by evolutionary response to disease. Even sexual attraction has a connection to disease. Why is the scent of someone you find sexually

attractive so alluring? It's often a sign that you have dissimilar immune systems, which will give your children wider immunity than
either of their parents.

Of course, it's not just external organisms we've evolved to
manage—or that have evolved to manage us. Guess what? You may
not have sent any invitations, but as you read this, you're playing
host or hostess to a massive party of microbes. In fact, if your body's
a party and your cells are the guests, you're outnumbered in your
own home. An adult human contains *ten times* as many "foreign"
microbial cells as mammalian cells. If you put them all together,
you'd find more than 1,000 different types of microbial creatures
weighing about three pounds and numbering somewhere between
10 trillion and 100 trillion. And when it comes to genetic material,
it's not even close; the microbes that make you their home col-
lectively contain 100 times as many genes as your own genome
does.

Most of these microbes are found in the digestive system, where
they play crucial roles. These intestinal bacteria, or gut flora, help
to create energy by breaking down food products we otherwise
couldn't break down; they help to train our immune systems to
identify and attack harmful organisms; they stimulate cell growth;
and they even protect us against harmful bacteria. In fact, the digestive problems many people experience when taking antibiotics
are directly related to the loss of these healthy bacteria. Using
broad-spectrum antibiotics is like carpet bombing—they kill
everything in their way and can't tell the difference between ene-
mies, allies, and innocent bystanders. That's why many doctors recommend eating yogurt when taking antibiotics: the bacteria in
yogurt are friendly—probiotic—and they can help to provide some

of the digestive assistance and protection that is normally per-formed by the gut flora until they get back to normal levels.

Not all the bacteria who have made you their home are so friendly—right now, you may be providing a human roof over the metaphoric heads of *Neisseria meningitidis, Staphylococcus aureus,* and *Streptococcus pneumoniae,* the bacteria that can cause, respec-tively, meningitis, toxic shock syndrome, and pneumonia. Fortu-nately, the millions of microscopic allies in your gut have also taken it upon themselves to keep the bad guys under control.

Through what's called the barrier effect, colonies of gut flora prevent these dangerous bacteria from growing to dangerous levels by dominating the resources in the digestive tract. The helpful bac-teria actually work with our own bodies to ensure that harmful bacteria can't gain a microscopic foothold. To provide a similar ef-fect, some doctors advise women who are prone to yeast infections to take probiotics, either by eating them in foods like yogurt or by taking a supplement. Just as they do in the digestive system, probi-otic friendly bacteria act as naturally occurring helpful bacteria and create a barrier effect that inhibits the growth of vaginal yeast. One of the reasons some probiotics are friendly has to do with their taste in metals. Remember how almost every form of life on earth needs iron to survive? Well, one of the exceptions is also one of the most common probiotics, a bacterium called *Lactobacillus,* which uses cobalt and manganese instead of iron—which means it's not hunting yours.

Your digestive system is a veritable jungle, with hundreds of species of bacteria competing for survival—most of them working with you, but a few of them ready to work against you if they have the chance. When the relationship between an organism and the host it inhabits is mutually beneficial—as is generally the case with humans and intestinal bacteria—it's called symbiotic. Often, of course, that's not the case. The Guinea worm is a pure parasite; it lives off its human host for its own benefit, providing nothing, causing only harm. And when its victim feels the natural urge to plunge the sores the worm causes into cool water (and thus help the worms to spread), the infected person is experiencing a type of host manipulation—the phenomenon that occurs when a parasite provokes its host to behave in a way that helps the parasite to survive and reproduce.

By examining some of the most extreme examples of host manipulation in nature, we can gain a better understanding of how parasites can affect our *own* behavior. So before we continue our exploration of the relationships among humans, microbes, and our mutual evolution, let's take a trip back to the actual jungle to examine a real-life *Invasion of the Body Snatchers,* Spider Body Snatchers, anyway.

PLESIOMETA ARGYRA IS an orb-weaving spider native to Central America. Orb weavers are a large family of spiders, with more than 2,500 different species spinning webs around the world. True to their name, these little guys spin those familiar circular webs with bull's-eye centers. The fellow we're concerned with—along with his special relationship to a parasitic wasp called *Hymeno-epimecis argyraphaga*—has been the subject of serious study by a scientist named William Eberhard. Because these insects only have Latin names, let's call the spider Thane of Cawdor and the wasp Lady Macbeth.

Cawdor lives a happy life in the Costa Rican jungle, spinning orb-shaped webs, hunting the prey that happen to stumble into his home, and wrapping them up for later consumption. Then one day Lady Macbeth flies up, seemingly out of nowhere, and stings him. Cawdor is paralyzed. Now the wasp lays an egg on the spider's abdomen. Ten to fifteen minutes later, Cawdor awakens and goes about his business—spinning webs and trapping prey. Little does he know that from the moment Lady Macbeth first laid her stinger on him he was as doomed as his namesake. The egg deposited by the adult wasp soon hatches into a larva. The larva—let's call it Baby Macbeth—makes holes in the spider's abdomen and slowly feeds off its blood. Over the next few days, the wasp larva lives off the spider and the spider spins on, oblivious.

Then, when the larva is ready to cocoon and begin the final phase of its transformation into an adult, Baby Macbeth injects old Cawdor with chemicals that completely change the spider's behavior, effectively turning it into the larva's slave. Instead of building circular webs, the spider now goes back and forth over the same few spokes—retracting its steps as many as forty times, as it builds a special web to protect the larva's cocoon. Then, near midnight (Mother Nature can definitely lay on the drama) the spider sits down in the center of this special web and doesn't move. All that's left is for Baby Macbeth to finish the job.

The larva kills the motionless spider and basically sucks it dry. When it's finished its meal, it discards the spider's lifeless husk on the jungle floor. The next night it spins a cocoon around itself, which it hangs from the reinforced webs built by the dead spider, and enters the final phase of its growth. Around a week and a half later, an adult wasp emerges from the cocoon.

Researchers aren't entirely sure how the larva hijacks the spider's instinctual web-building behavior. To be clear, it's not that the spider is behaving in a completely new and different way—the steps it repeats to build the special "cocoon web" are essentially the first two steps of the five basic steps involved in building a normal web; it just repeats them over and over again like some kind of looping music track stuck on repeat. Dr. Eberhard says, "The larva somehow biochemically manipulates the spider's nervous system causing it to perform one small piece of a subroutine, which is normally only a part of orb construction, while repressing all the other routines."

Dr. Eberhard's research also made it clear that, however the biochemical injected by the larva works, it works quickly and lasts awhile. In laboratory studies, when the parasite is removed from the spider *after* it has started to build the cocoon web but *before* it has finished—that is, after the larva has asserted mental control but before it kills the spider—our arachnid friend continues to build the cocoon web for days, until it eventually returns to building normal webs.

Nature abounds with examples of host manipulation; generally—no big surprise here—they involve a critical step in the parasite's efforts to reproduce. In the case of many parasites, that boils down to this—how do I get from this host to the next one? Before we turn back to parasites that manipulate humans, let's look at a parasite that faces a particularly vexing transportation problem.

DICROCOELIUM DENTRITICUM IS a tiny worm that lives in the livers of sheep and cattle; it's commonly called a lancet liver fluke. If you and your family lived in a sheep and you didn't want your entire species to die out when the sheep died, you'd have to find a way to get your kids into the gut of another sheep. When adult flukes lay eggs, those eggs are passed by their hosts in dung where they remain dormant until a land snail comes along to feed on the dung, eating the eggs in the process. Once eaten, the eggs hatch inside the snails, and, eventually, the newborn flukes are excreted from the snail as slime. Ants feed on the slime and become a new ride for the flukes in the process—but there's still a long road ahead. Think about it—you're riding in an ant and you need to get into a sheep; what to do?

As the worms being carried by the ant develop, one of them makes its way to the ant's brain, where it manipulates the ant's nervous system. Suddenly, the fluke-hosting ant behaves in completely uncharacteristic fashion. Every night, it leaves its colony, finds a nice blade of grass, and climbs to the tip, where it hangs on and, apparently suicidal, waits to be eaten by a grazing sheep as it munches on the grass. If it's not eaten, it returns to its colony during the day and finds another blade of grass the next night. Eventually, when the ant is eaten along with its blade of grass, the flukes make their way from the digestive system of their new host and colonize another liver.

The parasitic hairworm *Spinochordodes tellinii* grows to adulthood inside grasshoppers in the south of France. It's another worm that, like a houseguest that will never leave, makes its host suicidal. As soon as the hairworm larva reaches adulthood it releases specialized proteins that convince the unfortunate French grasshopper to find the nearest pool of water and jump right in, like a drunken sailor docked in Marseille who has forgotten that he can't swim. Once in the water, while the grasshopper is drowning, the worm slithers out and swims off to find romance and reproduction.

Remember, bugs and worms aren't the only organisms capable of host manipulation. Viruses and bacteria engage in sophisticated host manipulation all the time. The rabies virus is an interesting example of host manipulation on more than one level. The rabies virus colonizes the salivary glands of its host, making it difficult to swallow. That's what causes the characteristic foaming at the mouth—the inability to swallow makes the animal's mouth froth with, not coincidentally, rabies-filled saliva. By the time the animal is foaming at the mouth, the virus will most likely have infected its host's brain, where it chemically induces the animal to feel higher and higher levels of agitation and aggression. When animals are agitated and aggressive, they bite. When their mouths are foaming with rabies-filled saliva, their bites are infectious. Angry bite plus infected saliva equals new host, which means survival and reproduction for the virus. The origin of "foaming at the mouth" as an idiom for angry and aggressive behavior isn't the only piece of culture we've gotten from rabies. It's very likely that the werewolf myth, in which one bite transforms the victim into a possessed beast just like the biter, almost certainly has its roots in ancient observations of the rabies virus at work.

Enslaved spiders and suicidal grasshoppers are examples of host manipulation at its most extreme. Janice Moore, a professor of biology at Colorado State University who has studied host manipulation for more than twenty-five years, notes that, in some cases, the change can be so dramatic that the infected host is essentially transformed into another creature:

It is possible that the parasitized animals are frequently so altered compared with their uninfected counterparts that they well may be the functional equivalent of a different species.

On the other hand, many host manipulations are more subtle and at least *seem* to be natural. Notice, even in the case of the orb-weaving spider and the wasp larva, it's not that the larva actually assumes complete control of the spider. Rather, through chemical manipulation, it gets the spider to behave in a way that is more to the larva's benefit than to the spider's. But the spider is still alive and volitional—the two steps of the web-building routine, after all, belong to the spider, not the wasp. Similarly, when people infected with Guinea worm plunge their hands into a cold pool to relieve the pain, the Guinea worm isn't actually controlling their minds, of course—but it has evolved to stimulate its host to behave in a way that helps it survive and reproduce.

The good news for us is that we're a lot smarter than spiders. The more we understand how parasites manipulate their hosts, especially when their hosts are humans, the more we can manage those effects and control the outcome. Sometimes, the only effective option may be to stamp out the behavior that allows the threatening parasite to reproduce—as in the case of the Guinea worm. Sometimes, as you'll soon see, we may be able to steer the parasite's evolution in a more benign—or at least less harmful—direction. There's ample evidence of that in the evolutionary record, after all. Just think about all those bacteria in your stomach helping you to digest that pint of Häagen-Dazs you shouldn't have eaten for lunch.

TOXOPLASMA GONDII IS a parasite that can infect just about every warm-blooded animal but can reproduce in a way that guarantees its survival only in cats. *T. gondii* reproduces by copying itself during the life of its host, but it's only in cats that it undergoes sexual reproduction, producing new oocysts, or spore cells, that can go on to find new hosts. Infected cats distribute oocysts in their droppings. The oocysts are hardy little organisms that can survive for as long as a year in tough conditions. When rodents, birds, or other animals ingest the oocysts, they become infected; animals can also become infected by eating the flesh of an infected animal. Humans can ingest oocysts by eating undercooked meat or poorly washed vegetables or after handling cat litter.

Once an animal is infected, the *T. gondii* cells are distributed through the body by the bloodstream, where they insert themselves inside muscle and brain cells. It's a pretty nasty-sounding infection—who wants parasites setting up permanent shop in your brain?—but it's thought to be generally benign in most people, although more on that shortly. It's also incredibly common, infecting as much as half the world's people—and not just where you might think. According to the Centers for Disease Control and Prevention (CDC) in the United States, scientists think more than 20 percent of the population is infected—in France, it's nearly 90 percent. (Some epidemiologists think there's a correlation between raw meat consumption and *T. gondii* infection rates, which might somewhat explain the high level of French infection; *tartare* is a French word, after all.)

None of which explains how *T. gondii* gets back into a cat. Well, that's where the story gets interesting. *T. gondii* is a master little host manipulator—of mice and rats. When a mouse (or a rat) eats infected cat droppings, the parasite behaves in the usual manner, moving into the mouse's muscle and brain cells. Once inside the mouse's brain, in ways that are not completely understood, the parasite has a profound effect on its behavior. First, the mouse becomes fat and lethargic. Then, it loses its natural fear of predators—of cats.

In fact, some studies have shown that instead of fleeing areas marked with cat urine, infected mice are actually drawn by its scent. You know what the scientific term is for a fat, slow mouse that is attracted by the smell of cats?

Cat food.

Which gets *T. gondii* exactly where it wants to go.

We mentioned a moment ago that *T. gondii* is thought to be largely benign in humans. Well, that is *largely* the case, but not always the case. First of all, people with severely compromised immune systems, like people living with HIV, are at risk for serious complications, as they are with many infections that people with a fully functioning immune system can manage. Those complications can include blindness, damage to the heart and liver, and inflammation of the brain, called encephalitis, which can lead to death. The other group that has to be on the lookout is pregnant women. Depending upon how far along she is, if a pregnant woman becomes infected, there can be as much as a 40 percent chance the fetus will become infected, and that can cause similar severe complications. This risk doesn't exist if a woman is *already* infected, that is, if she became infected at some point before she became pregnant—there's only a risk to the fetus during the phase of initial infection. But for that reason, pregnant women and people who have compromised immune systems should avoid raw meat and let somebody else empty the litter box.

There is also increasing evidence that past infection with *T. gondii* (toxoplasmosis) may trigger schizophrenia in some people. E. Fuller Torrey, a renowned psychiatrist and schizophrenia researcher, publicized many of these theories in 2003. It seems clear that there is a higher incidence of *T. gondii* infections in schizophrenics—although it isn't yet clear what causes what.

T. gondii may be a schizophrenia trigger, but it's also possible that people with schizophrenia are more likely to engage in behavior that exposes them to *T. gondii*, like poor hygiene. It's certainly an area that deserves serious exploration—just a decade ago, scientists dismissed the idea that infections could cause ulcers; today that's a proven fact. (Of course, the doctor who proved the connection, Dr. Barry Marshall, had to swallow bacteria and give himself an ulcer before the "experts" would take it seriously. Sometimes there is justice, though; Dr. Marshall along with his colleague J. Robin Warren won the Nobel Prize in physiology or medicine in 2005 for their discovery.)

The notion that *T. gondii* may trigger schizophrenia is supported by recent studies demonstrating that mice that have toxoplasmosis modify their behavior when given antipsychotic medication. Researchers at Johns Hopkins University are now testing whether schizophrenics might be helped with antibiotics that fight toxoplasmosis. If Dr. Torrey is right, and *T. gondii* infection *can* trigger schizophrenia, it will add a whole new meaning to the stereotypical picture of the crazy cat lady.

Given *T. gondii*'s dramatic influence on rodent brain chemistry, it's not surprising that scientists have looked for evidence that the parasite influences humans as well. And there is evidence that people who have *T. gondii* infections do exhibit some subtle differences in behavior when compared to uninfected people. Again, it's not clear whether *T. gondii* is *causing* the behavior or whether people with these behavioral tendencies are more likely to be exposed to *T. gondii*—but it *is* interesting.

One dedicated researcher, Professor Jaroslav Flegr of Charles University in Prague, has discovered that women infected with *T. gondii* spend more money on clothes and are consistently rated as being more attractive than women without the infection. Flegr summed up his findings this way:

> We found they [infected women] were more easy-going, more warm-hearted, had more friends and cared more about how they looked. However, they were also less trustworthy and had more relationships with men.

Flegr found infected men, on the other hand, to be less well groomed, more likely to be loners, and more willing to fight. They were also more likely to be suspicious and jealous and less willing to follow rules.

If it turns out that *T. gondii* does influence human behavior in any of these ways, it's likely to be an accidental effect of the parasite's evolved manipulation of rodents. That's part of the reason why the possible effects in humans seem so much subtler than the effect in rodents—the manipulation is designed to get rodents to be eaten by cats, because that's where *T. gondii's* primary life cycle occurs. The infection of humans and other animals is more or less gravy for the parasite. The chemicals *T. gondii* evolved in order to affect the behavior of rodents may also have an effect on our brains. But whatever effect they do have isn't host manipulation in the evolutionary sense, because it doesn't do anything for the parasite—unless you know about a species of cats that only eats well-dressed women.

MOST PEOPLE THINK of sneezes as symptoms—but that's really only half the story. A normal sneeze occurs when the body's self-defense system senses a foreign invader trying to get in through your nasal passages and acts to repel the invasion by expelling it with a sneeze. But sneezing when you've got a cold? There's obviously no way to expel the cold virus when it's already lodged in your upper respiratory tract. That sneeze is a whole different animal—the cold virus has learned to trigger the sneezing reflex so it can find new places to live by infecting your family, your colleagues, and your friends.

So yeah, sneezes are symptoms—but when they're caused by a cold, they're symptoms with a *purpose,* and the purpose isn't yours. That's true for many of the things we think of as symptoms of infectious disease—they're actually the product of host manipulation as whatever bacteria or virus has infected us works to engage our unconscious assistance in making the jump to its next host.

As many people who have children know, pinworm infection is one of the most common infections contracted by children in North America. The CDC believes that somewhere around 50 percent of American kids probably have pinworms at any given point in time. Adult pinworms are no more than half an inch long and look more or less like a small piece of white thread. Pinworms grow to maturity in the large intestine, where they feed on digestive matter and eventually mate. During the night, pregnant females make their way out of the large intestine (the same way everything else does) and deposit their microscopic eggs on the skin of the infected child. At the same time, they deposit allergens that cause serious itching. They don't usually cause any damage except the itching—but those worms definitely want your child to scratch that itch.

When a child who has pinworms scratches his or her bottom, the eggs get lodged underneath his or her fingernails. Without serious scrubbing every morning, including underneath fingernails, it's easy for those eggs to get around. They're sticky little things and they easily make their way from fingers to everything the child touches—doorknobs, furniture, toys, even food. When other children touch those surfaces, they pick up some eggs. Eventually, those curious fingers make their way into mouths and some eggs are ingested orally, worms hatch in the small intestine, migrate to the large intestine, and begin the cycle again. Pinworms live only in humans—contrary to popular belief, they can't be caught from any other animal (although their eggs could easily be picked up from the fur of a pet that had been touched by a person with eggs on his or her fingers). Their survival requires movement from human host to human host, and they've evolved a simple and efficient method of host manipulation to help them make the trip—scratch and spread.

Other diseases cause symptoms that manipulate us in more passive ways, all in the name of easing their ability to spread and reproduce. Cholera is a waterborne disease that causes severe diarrhea. In serious cases, the persistent diarrhea can cause dehydration and death. But like the itching caused by pinworms and the sneeze caused by the cold, the diarrhea caused by cholera isn't just a symptom—it's a transmission channel. It's how the disease makes it into the water supply and ensures its ability to find new hosts.

Malaria manipulates human hosts too—in its case, by incapacitating us. People with malaria experience a terrible cycle of fever and chills, accompanied by debilitating weakness and fatigue—and when you're lying in bed too tired even to lift an arm, you're a pretty helpless target for mosquitoes. Mosquitoes bite infected humans and pick up a load of malaria-causing protozoa, and then these bugs carrying bugs fly away to infect someone else.

The study of host manipulation in humans is very young, but

it's already revealing some surprising insights that promise new insights into the causes—and potential cures—of an enormous range of disease. We've discussed the possibility that when *T. gondii* jumps from cats to cat owners, it may sometimes trigger schizophrenia. Recent, although controversial, research shows the possibility of a connection between obsessive-compulsive disorders and streptococcal infections in children.

The family of streptococcal bacteria is responsible for a wide range of human disease—from strep throat to scarlet fever, bacterial pneumonia, and rheumatic fever. Many types of streptococcal bacteria exhibit a phenomenon called molecular mimicry in which they display characteristics of human cells in order to trick the immune system. The cells these bacteria mimic include cells found in the heart, the joints, and even the brain. When you have a bacterial infection, your immune system produces antibodies to attack the invaders. When the invaders are partially disguised through molecular mimicry, they can cause an autoimmune disorder. The immune system recognizes the threat posed by bacterial invaders, but the antibodies it produces attack all the cells that resemble the bacteria—including the body's own cells. That's how some children who have rheumatic fever end up with heart problems—antibodies attack the heart valve because the infecting bacteria resembles it in some ways.

Dr. Susan Swedo, a researcher at the National Institute of Mental Health, believes that certain strep infections can trigger an autoimmune disorder that leads to an antibody-led attack on the basal ganglia, the part of the brain believed to control movement. Researchers call this condition PANDAS—pediatric autoimmune neuropsychiatric disorder associated with streptococcal infection.

Parents of children with PANDAS describe heartbreaking trans-
formations, often overnight. Shortly after infection, children suddenly display repetitive tics and uncontrolled touching, as well as
serious anxiety.

It's not clear that this is actual host manipulation—that depends on whether the change in behavior helps the bacteria to
spread. Theoretically, of course, it's not hard to imagine how uncontrolled, repetitive touching of toys, furniture, and other kids
would help the virus to spread. It's also possible that there is a relationship between obsessive-compulsive disorder and strep infections that isn't host manipulation itself, but the by-product of the
bacteria's effort to fool the immune system.

One thing is clear—we are just beginning to understand the
myriad ways our behavior is affected by infectious agents. One very
new avenue of research is exploring the striking possibility that
sexually transmitted diseases may actually influence sexual behavior. Now, I'm not suggesting that this kind of influence will trans-
form a happily married man into an insatiable cheat. In fact, that
wouldn't necessarily be in the virus's (or fungus's or bacteria's) interest. Too much promiscuity on the part of the host could disable
it with other, potentially more damaging, diseases. And that would
leave the parasite stuck in a host that couldn't get around. From the
sexually transmitted parasite's point of view, it may want you to
have more sex—but not *too much* sex.

As far as diseases influencing human sexual behavior, some researchers are examining the possibility that genital herpes may affect human sexual feeling in a way that could influence behavior.
Two researchers at the Department of Anatomy and Neurobiol-
ogy at the University of California at Irvine, Carolyn G. Hatalski

and W. Ian Lipkin, have speculated that the herpes virus may heighten sexual feeling because it is so intertwined with the nerves that carry those feelings. They wrote:

> It is intriguing to speculate that the ganglion infection may modulate sensory input to sex organs leading to increased sexual activity and enhanced probability of virus transmission.

In other words, sometimes the herpes virus may want you to get some action.

HOST MANIPULATION OCCURS when a parasite or disease affects our behavior for its own ends. But that's not the only way disease affects human behavior, of course—there are thousands of ways in which personal, cultural, and social standards have evolved in order to help us avoid or manage disease. Some behavior is instinctual, like the sense of disgust at certain sights and smells, which prompts us to avoid animal waste or spoiled food—things that are usually ripe with infectious material. Others are learned behavior and social pressure—covering your nose and mouth when you sneeze is a good example. Washing your hands before a meal is another. All of these reactions to disease are called behavioral phenotypes—the observable actions of an organism that result from its attempts to manage the interaction between its genetic makeup and its environment for its own benefit.

A few evolutionary psychiatrists (scientists who study human behavior in the context of evolution and look to see whether specific behavior conferred an evolutionary advantage) have even suggested that humankind's instinctual fear of strangers may have its roots in disease avoidance. The theory is rooted in the notion that in humans two of our basic biological imperatives—survival and reproduction—have fostered in us a core social concern for the health and safety of our children and close relatives. This concern means that, in certain circumstances, evolution might actually push us to sacrifice our own survival for the sake of our children's survival, or even that of close relatives. And the more relatives you could save through your sacrifice, so the theory goes, the more likely you'd be to act. From an evolutionary perspective, it makes perfect sense—let a single carrier of your genes die (that is, you) in order to let your larger gene pool of close relatives and extended family survive.

So what happens when you're sick with a deadly—and contagious—infection? Some researchers believe that the sick primate that is abandoned by its community may actually be partly responsible, wandering away to protect its kin from infection. This phenomenon has been documented in cliff swallows and flour beetles; when they're infected with parasites, members of both species appear to migrate away from their kin.

There's also evidence that some species have evolved mechanisms to avoid their brethren when they become infected with a dangerous parasite. Researchers at Old Dominion University, in Norfolk, Virginia, studied Caribbean spiny lobsters, usually gregarious critters that normally live together in communal dens. The researchers found that when otherwise healthy lobsters become infected with a lethal pathogenic virus they are shunned by their den mates—the uninfected lobsters pick up and leave. What's really amazing is that the uninfected lobsters make for the underwater highway before the diseased lobster shows any symptoms. Which means the behavior is likely to involve some chemical sensor and trigger.

Here's where it all comes together as far as this theory is concerned. If certain infections drive organisms away from their own group in order to protect their kin, how will other groups respond when an unknown individual comes wandering over the hill? Xenophobia, which is the formal name for the fear of outsiders, appears to be a nearly universal instinct in human culture. It's possible that xenophobia has its roots in some deeply buried instinct to protect one's own group from outside threats to health and survival, including infectious disease. Of course, if that *is* the case, understanding its origins will give us another powerful tool in fighting an instinct—if it even is one—that has long outlived its usefulness.

" 'SUPERBUGS' SPREAD FEAR FAR AND WIDE"

"RISING DEADLY INFECTIONS PUZZLE EXPERTS"

"BACTERIA RUN WILD, DEFYING ANTIBIOTICS"

You've seen the headlines. They've probably frightened you. And it's true—just as we've been evolving to survive disease, all the organisms that cause disease have been evolving right along with us. You've seen how parasites have evolved very specialized abilities to navigate seemingly impossible challenges to survival—like traveling from a sheep to a snail to an ant in order to get to another sheep. And small organisms, because they multiply so rapidly and so frequently, sometimes cycling through hundreds of generations in just days, have one big evolutionary advantage over us—they evolve faster. Take *Staphylococcus aureus*, which doctors call staph for short. Staph is a very common bacteria; it may be living on your skin or in your nose right now. It can cause pimples—and it can cause deadly infections like meningitis and toxic shock syndrome. It's also the bug behind many of those terrifying reports of antibiotic-resistant infections plaguing hospitals and, more recently, professional and college sports teams.

When penicillin was accidentally discovered by Alexander Fleming in 1928, it was actually inhibiting the growth of staph—that's what was in the petri dish. Fourteen years later, when penicillin was first used to treat infections in humans, there were virtually no reports of penicillin-resistant staph. But just eight years later, in 1950, 40 percent of all staph infections were penicillin-resistant. By 1960, that number had climbed to 80 percent. Treatment switched to a specialized relative of penicillin called methicillin, which was introduced in 1959—and two years later, the first incident of methicillin-resistant staph, known as MRSA, was reported. MRSA is now firmly entrenched in hospitals, and treatment has moved to a different class of antibiotics, usually with one called vancomycin. The first case of VRSA—yes, vancomycin-resistant staph—was reported in 1996 in Japan.

All of this sounds frightening—as if we're in an arms race where the other side has vastly superior technology. But that's not the whole story—they're faster, but *we're smarter*. We can think about how evolution works and try to use that to our advantage—they can't think at all. Now, remember that the biological imperatives driving bacteria are survival and reproduction, just like the biological imperatives that drive everything else. So what if we made it easier for a given type of bacteria to survive in a *healthy* human than to survive in a sick human—wouldn't that create evolutionary pressure against behavior that harms us?

That's what Paul Ewald thinks.

PAUL EWALD IS one of the pioneers of evolutionary biology, especially the evolution of infectious diseases and how pathogens select for—or against—traits that harm their hosts. The degree to which an organism destroys its host is called virulence. The range of virulence found in pathogens that infect humans is enormous—from all-but-harmless (pinworms) to unpleasant but hardly dangerous (the common cold) to rapidly, horribly fatal (Ebola). So why does one microbe evolve toward massive virulence while another is content to leave you up and running? Ewald believes the key factor that determines virulence is how a given parasite gets from host to host.

When you remember that every infectious agent has the same goal—to survive and reproduce by infecting new hosts—that starts to make a lot of sense. Let's look at the three basic ways a microbe moves from one host to another:

- Close proximity that allows for transmission through the air or physical contact—diseases transmitted this way include the common colds and sexually transmitted diseases (STDs)

- Hitching a ride on an intermediate organism, usually a mosquito, fly, or flea—this category includes malaria, African sleeping sickness, and typhus

- Traveling through contaminated food or water—cholera, typhoid fever, and hepatitis A are all transmitted this way

Now let's think about what that means in terms of virulence. According to Ewald, diseases in the first category face evolutionary pressure *against* virulence. These microbes rely on their hosts to carry them around and introduce them to new hosts. That means they need their hosts to be relatively healthy—certainly healthy enough to be mobile. That's why you can almost always get up and go to work when you've got a cold, even if you're miserable the whole time. The cold virus leaves you well enough to get on the subway and go to work, sneezing and coughing all the way. Ewald believes the cold virus has hit the evolutionary jackpot; it's evolved to a level of virulence that guarantees our mobility and its survival. In fact, he believes it may *never* evolve to kill or seriously incapacitate us.

On the other hand, when an infectious agent doesn't need its host to get around, things can really heat up. As we mentioned, malaria has evolved to incapacitate us—it doesn't need our help to meet new hosts; instead, it wants us vulnerable to attack by its blood-sucking buddies, mosquitoes. In fact, there is an evolutionary *advantage* for the malaria parasite to push its hosts toward the brink of death. The more parasites swarming through our blood, the more parasites the mosquito is likely to ingest; the more parasites the mosquito ingests, the more likely it will cause an infection when it bites someone else.

Cholera is similar—it doesn't need us moving around to find new hosts, so there's no reason for the bacteria to select against virulence. It spreads easily through unprotected water supplies when soiled clothes or bed linens are washed in rivers, ponds, and lakes, or through sewage runoff. And again, cholera actually has an advantage in evolving toward virulence—as the bacteria reproduces ruthlessly, causing more and more diarrhea, the infected person may excrete as many as a billion copies of the organism, increasing the likelihood that some bacteria finds its way to a new host.

The bottom line is this—if an infectious agent has allies (such as mosquitoes) or good delivery systems (such as unprotected water supplies), peaceful coexistence with its host becomes a lot less important. In those cases, evolution is likely to favor versions of the parasite that best exploit its host's resources, allowing the parasite to multiply as much as possible—all of which spells bad news for the host.

But not necessarily bad news for humanity: Ewald believes that we can use this understanding to influence the evolution of parasites away from virulence. The basic theory is this—shut down the modes of transmission that don't require human participation and suddenly all the evolutionary pressure is directed at allowing the human host to get up and get out.

Let's look at how this would apply to a cholera outbreak. According to Ewald's theory, the virulence of a cholera outbreak in a given population should be directly related to the quality and safety of that population's water supply. If sewage flows easily into rivers that people wash in or drink from, then the cholera strain would evolve toward virulence—it can multiply freely, essentially using up its hosts, relying on its access to the water supply for transmission. But if the water supply is well protected, the organism should evolve away from virulence—the longer it remains in a more mobile host, the better its chance of transmission.

A series of cholera outbreaks that began in Peru in 1991 and spread across South and Central America over the next few years provide compelling evidence that Ewald is on to something. The water supply systems from country to country ranged from relatively advanced to seriously rudimentary. Sure enough, when the bacteria invaded nations with poorly protected water supplies, such as Ecuador, the virus became more harmful as it spread. But in countries with safe water supplies, such as Chile, the bacteria evolved downward in virulence and killed fewer people.

The implications of this are huge—instead of challenging bacteria to become stronger and more dangerous through an antibiotic arms race, we could essentially challenge them to get along with us. Think about the application of this theory just in terms of waterborne diseases like cholera. If we clean up water supplies, it will certainly mean fewer people will get infected because fewer people will consume contaminated water. But if Ewald is right, every dollar spent on protecting water supplies—and thus, controlling the transmission channel of the disease—will also steer the evolution of the disease itself toward a less harmful incarnation. As Ewald said:

> We should be taking control of the evolution of those disease organisms, favoring those mild strains and thereby essentially domesticating those disease organisms, making them into mild versions of what was there before. With a mild version, most people won't even know they're infected. It'll be almost like those people having a free, live vaccine.

If every malaria patient were covered in mosquito netting or stayed indoors, we might push *P. falciparum,* the malaria-causing protozoa, in a similar direction. If mosquitoes didn't have access to bedridden malaria patients, the microbe would be under evolutionary pressure to evolve in a way that allowed the infected person to remain mobile, increasing the opportunity for it to spread.

Of course, Ewald knows that his theory isn't always applicable.

Some parasites complicate the picture because they are capable of survival outside a host for a very long time. A pathogen that can lie in wait for years until a potential host happens upon it isn't very reliant on transmission pressure. Anthrax is one of these patient predators. The deadly bacteria can exist outside a host for more than ten years in some situations. In these cases, it's hard to affect virulence by reducing the pathogen's transmission channels, because its ability to survive outside a host makes it less concerned with transmission from an evolutionary perspective.

WE ALREADY KNOW that humans can affect the evolution of bacteria. The evolution of all those antibiotic-resistant strains of staph is conclusive proof of that. But Ewald's theory takes the notion that bacterial evolution gives *bacteria* an advantage over us and turns it on its head:

> Not by getting involved in some kind of arms race in which we're using one antibiotic weapon against the organism, and [the organism] evolve[s] a defensive weapon against that antibiotic, and then we have to shift to another, and so on, indefinitely. Instead, we have a sense of where we want evolution to end, and we adjust the environment so that the organism freely evolves to that endpoint, which is in its interest and also in our interest.

By understanding how the organisms that cause infectious disease have evolved among us, next to us, and inside us—affecting their evolution even as they affect ours—we gain new insight into how those diseases influence us, and into how they can be controlled for our benefit. Already, that understanding is giving us the opportunity to interrupt the transmission channel of horrible afflictions like the Guinea worm. And it suggests powerful ways to change the course of diseases—like cholera and malaria—that have plagued humankind for longer than there has been a history to record it.

When it comes down to it, everything that's alive wants to do two things: survive and reproduce. The Guinea worm wants to, the malaria protozoa wants to, the cholera bacteria wants to—and so, of course, do we. The difference—our big advantage—comes down to one thing.

We know it.

CHAPTER VI

JUMP INTO THE GENE POOL

E dward Jenner was just a country doctor in Gloucestershire, England, at the end of the eighteenth century when he noticed a surprising pattern. Milkmaids who caught cowpox (that's what happens when you used to spend a lot of time with cows), a very mild infection in humans, seemed to be resistant to smallpox, a very deadly infection in humans. So Jenner wondered whether he could duplicate the effect intentionally. He scraped a cowpox sore on an infected milkmaid and purposefully infected several teenage boys. Sure enough, his hunch was correct. The cowpox infection resulted in protection from smallpox, and Edward Jenner—not just a simple country doctor after all—had the first vaccine on his hands. The word *vaccine* actually comes from the Latin word for cow, *vacca,* and the Latin name for cowpox, *vaccinia.*

Today, we know a lot more about how vaccination works. It begins with a relatively harmless version of the virus we want to vaccinate against (harmless because it's weakened or killed and broken up into pieces or, like cowpox, a relative close enough to the harmful virus that our bodies will recognize it, but distant enough that it does not cause serious disease). By introducing the harmless virus to our bodies, we stimulate our immune systems to produce antibodies specifically tailored to defend against that virus. Then, if we *are* exposed to the harmful version, our bodies are prepared to defend themselves immediately. Cowpox, for example, causes only a very mild infection in people, but its structure is so close to that of smallpox that the antibodies our immune systems produce to fight cowpox will also work against smallpox. Without having the right-fitting, preformed antibodies, viral attackers can make us sick before our immune system has time to generate the antibodies we need to fight back.

Now, here's where it really gets interesting. There's a massive number of potential microbial attackers out there, and our bodies produce a specific antibody to fight back against each and every one. For a long time, scientists couldn't understand how that worked—there just didn't seem to be enough active genes in humans to direct the production of all these antibodies.

Of course, they didn't know that genes could *change*.

EVERY HUMAN BEING starts off with exactly the same number of cells as the simplest form of bacteria—one. That single cell, or zygote, is the product of the union of two other cells—a sperm cell supplied by the father and an egg cell supplied by the mother—that combine to produce a human in progress. Millions of years of evolutionary pressure, response, adaptation, and selection come together in that first cell—it contains every single genetic instruction to manufacture the proteins used to build a human being. All of those instructions are carried in about 3 billion pairs of nucleotides; those pairs of nucleotides are called DNA base pairs, of which there are assumed to be less than 30,000 genes. The genes themselves are organized among twenty-three pairs of chromosomes, for a total of forty-six.

One set of twenty-three chromosomes comes from the mother and one comes from the father. Every pair except for the twenty-third—the sex chromosomes—is a matched pair. In other words, each chromosome carries the same type of instructions, although they will vary greatly in how they instruct your body to carry out those instructions. For example, let's just say that specific chromosomes contain instructions for whether or not you'll have hair on your fingers; the instructions may code for hairy fingers in the chromosomes that come from your father, while coding for hairless fingers in the chromosomes that come from your mother. In that case, you will have hair on your fingers—the trait for hairy fingers is dominant, while the trait for hairless fingers is recessive. That means one copy of the fictious gene for hairy fingers is enough to ensure that you exhibit that trait. You need two copies of the gene for hairless fingers—one from your mother and one from your father—in order to have hairless fingers yourself.

Usually, with one very important exception, every cell in your body contains the same DNA—two complete sets of chromosomes with all the genes containing all the instructions you need to build every type of protein and every type of cell. The exception is germ cells, the cells that combine to produce offspring. Sperm and eggs each contain only one set of twenty-three chromosomes; when they unite to form a zygote, the resulting cell has a full complement of forty-six chromosomes, two sets of twenty-three each. But from the moment you're a sparkle in daddy's eye and a single-celled zygote on the way to implanting in mommy's uterus, every other cell includes your complete blueprint. Your toenails have the code to build brain cells—and your brain cells have the code for toenails. And fingernails. And blood cells. And just about everything else in your body.

But what's even more interesting is that less than 3 percent of your DNA contains instructions for building cells. The vast majority of your DNA—97 percent of it—isn't active in building anything. Think about that. If you took the DNA from any cell in your body and laid it end to end, it would reach the top of Shaquille O'Neal's head—but the DNA that actively codes for building your body wouldn't even reach his ankle.

Scientists initially called all this additional genetic material "junk DNA." They originally assumed that if it didn't code for cellular production, it was essentially parasitic—more or less lounging in the gene pool for millions of years without making any contribution to its upkeep. In other words, they thought this DNA did nothing for us at all; they imagined it was just hitching a ride through life, not hurting us, not helping us, just helping itself.

A series of new research is beginning to demonstrate that the previous assumption that so-called junk DNA is junk—was bunk. It turns out that the massive volume of genetic information in this portion of our genome may play a critical role in evolution. As its importance has been reevaluated, the respect it gets from the scientific community has begun to change; the standard term for this genetic material has even been upgraded—from junk DNA to noncoding DNA, which means it isn't directly responsible for making proteins.

Perhaps the biggest surprise is where much of this noncoding DNA comes from. You know that idea of a blissful future when bacteria, viruses, and humans live together in happy, healthy coexistence? What if I told you it's already sort of happening?

Almost every human cell contains microscopic workhorses called mitochondria that function as dedicated power plants, producing the energy to run cells. Most scientists now believe that mitochondria were once independent, parasitic bacteria that evolved a mutually beneficial relationship with some of our pre-mammal evolutionary predecessors. Not only do these likely former bacteria live in almost all your cells, they even have their own inheritable DNA, called mitochondrial DNA, or mtDNA.

Former bacteria aren't the only microbes we've married. Researchers now believe that as much as a third of your DNA is from viruses. In other words, our evolution hasn't only been shaped by *adaptation* to viruses and bacteria—it's probably been shaped by *integration* of viruses and bacteria.

UNTIL RECENTLY, THE scientific community all but universally agreed that genetic changes were the product of accidental mutations, caused by errors that were only random and always rare. Here's how those mutations happen. When cells are produced, DNA is copied from the "parent" cell to the "daughter" cell. This process usually produces accurate copies, but errors in the production of the long string of information that composes DNA do occur. In order to protect an organism against these errors, the transcription process is complemented by a proofreading system. Those proofreaders are so good that if we cloned them for publishers, they'd put copy editors out of business. Their error rate is phenomenally low—just one out-of-place nucleotide in every billion copies. When an error does get through, that new combination of DNA sequences, however slight, is a mutation.

Mutations also occur when organisms are exposed to radiation or powerful chemicals (like those found in cigarette smoke and other carcinogens). When that happens, it can also rearrange DNA. Before genetic engineering enabled us to modify food on a molecular level, plant breeders who wanted to create more efficient crops (hardier or more fruit-bearing, for example) would irradiate seeds by blasting them with a ray gun that could have come straight out of *Star Trek,* and then hope for the best. Most of the time, seeds couldn't even sprout after being irradiated, but every once in a while this heavy-handed genetic manipulation produced a beneficial trait.

Even the sun can cause mutation—not just by frying your skin and causing skin cancer, but on a global scale. Every eleven years, sunspot activity peaks and increased solar radiation explodes from the sun. Much of that energy is deflected by the earth's gigantic magnetic field, but some of it can "leak" through and play havoc.

In March 1989, a peak in sunspot activity led to a huge power surge that left more than 6 million people without power in parts of the northeastern United States and Canada. The sun spewed out so much energy that satellites were knocked out of orbit, garage doors began to open and close in California, and millions of people were treated to a version of the northern lights in places as far south as Cuba.

That may not be all the havoc these sunspot peaks cause. There's a curious correlation between these sunspot peaks and flu epidemics. In the twentieth century, six of the nine sunspot peaks occurred in tandem with massive flu outbreaks. In fact, the worst outbreaks of the century, killing millions in 1918 and 1919, followed a sun-spot peak in 1917. This might just be coincidence, of course.

Or it might not. Outbreaks and pandemics are thought to be caused by *antigenic drift*, when a mutation occurs in the DNA of a virus, or *antigenic shift*, when a virus acquires new genes from a related strain. When the antigenic drift or shift in a virus is significant enough, our bodies don't recognize it and have no antibodies to fight it—and that spells trouble. It's like a criminal on the run taking on a whole new identity so his pursuers can't recognize him. What causes antigenic drift? Mutations, which can be caused by radiation. Which is what the sun spews forth in significantly greater than normal amounts every eleven years.

The potential for evolution begins when a mutation occurs during the reproductive process of a given organism. In most cases, that mutation will have a harmful effect or no effect at all. Rarely, a random mutation will confer an advantage on its carrier, giving it a better chance to survive, thrive, and reproduce. In those cases, natural selection comes into play, the mutation spreads throughout the population through successive generations, and you have evolution. Adaptations that confer truly significant benefit to a species will eventually spread across an entire species, as when a strain of the flu virus acquires the new characteristic to go pandemic. But organisms, so the collective wisdom went, only happen upon helpful mutations by chance. (Remember, of course, that one species' advantage may be another species' disadvantage—an adaptation that allows a bacterium that harms humans to resist antibiotics is an advantage for the bacteria; for us, not so much.)

According to this way of thinking, the genome of every creature, great and small, lacks any ability to react intentionally on a genetic level to environmental changes that threaten its ability to survive and reproduce. It has to depend on luck to find a helpful mutation, or so the thinking goes. When the common strep infection evolves a trait that gives it antibiotic resistance, it's all luck. When humans evolved to cope with the rapid onset of the Younger Dryas, it was all luck. To be clear, scientists thought natural selection was influenced by the environment—but mutation never was. Mutation was an accident; natural selection occurred when the accident was helpful.

The problem with this theory is that it takes the evolution out of evolution. After all, what would be a more helpful mutation than one that allowed the genome to react to environmental changes and pass on helpful adaptations to successive generations? Surely, evolution would favor a mutation that helped an organism to discover adaptations that would help it survive. Saying otherwise is like saying that the only part of life not subject to evolutionary pressure is evolution itself.

The only-random-changes theory looks even weaker in light of recent work to map the human genome. Geneticists originally believed that every single gene had a single purpose—a gene for eye color, a gene for a widow's peak, a gene for attached earlobes. When genes went wrong, you ended up with a gene for cystic fibrosis, a gene for hemochromatosis, a gene for favism. That theory suggested the existence of more than 100,000 genes. But today, because of all the work that's gone into genome mapping, the total number of genes is thought to be about 25,000.

Suddenly, it's clear that genes don't have discrete jobs at all—there wouldn't be nearly enough genes to produce all the proteins necessary for human life if each gene only had one job. Instead, single genes have the capacity to produce many, many different proteins through a complex process of copying, cutting, and combining instructions. In fact, like a casino dealer who never stops, genes can shuffle and reshuffle endlessly to produce a huge array of proteins. There's one gene in a type of fruit fly that can produce almost 40,000 different proteins!

All this shuffling isn't restricted to single genes, either—the genetic dealer can borrow cards from other decks, combining parts of one gene with another. On a genomic level, that's where most of the complexity lies—and it's where the genetic work of making us human really happens. We may have the same genes as many other organisms, but it's what we do with them that counts. Of course, the idea that our genome can change has suddenly blurred the lines of what precisely a gene actually is. Yet, from an efficiency perspective, it makes a lot of sense for genes to be resourceful and to maximally utilize existing genetic parts. It's similar to the Japanese managerial system Kaizen, made famous in the 1980s. According to Kaizen, many working decisions are made on the factory floor and then communicated up to management—it's much more efficient to make a minor modification to an assembly line than to redesign the whole line.

On top of that, there are all kinds of redundancies built into the system. Scientists discovered this when they isolated specific genes related to specific functions in some organisms and removed those genes. They were shocked when these "knockout" (KO) experiments often did nothing at all; removing the gene in question simply had no effect. Other genes essentially stepped up and filled in for their KOed colleague.

Instead of imagining genes as a set of discrete instructions, scientists have begun to conceive of them as an intricate network of information, with an overall regulatory structure that can react to change. Like a foreman at a construction site who directs a particularly fast welder to pick up the slack when his buddy doesn't show up for work, the genome system can react to a knocked-out gene and get a body built just the same. Except the foreman isn't only a particular gene giving orders; rather, the whole system is interconnected and automatically covers for its parts.

You can see how these discoveries make it even harder to imagine how evolution relied only on random little changes in the code of individual genes to find the myriad adaptations that have allowed every living thing on earth to survive. If removing whole genes often has no effect on a creature, how could such minor changes be the only chance for the evolution of a new species, or even the successful adaptation of an existing one?

They probably can't.

JEAN-BAPTISTE LAMARCK WAS a French thinker and student of nature who popularized some of the current thinking about evolution and heredity in 1809 with the publication of his book *Zoological Philosophy*. In popular accounts of the history of evolutionary theory, Lamarck is built up into a somewhat foolish scientist who advances a series of wrongheaded theories about evolution and eventually "loses" an intellectual war with Charles Darwin.

According to the popular story, Lamarck was the chief proponent of a theory of inherited acquired traits. The essence of that theory is the idea that traits acquired by a parent during his or her lifetime could then be passed on to his or her offspring. It's suggested, for example, that Lamarck believed that giraffes' long necks were the result of each generation's straining its neck ever farther to reach leaves on higher branches. Or that a blacksmith's son would be born with stronger arms because his father developed those muscles hammering against his anvil. According to the myth about Lamarck, Darwin came along and proved Lamarck all wrong, debunking the notion that traits acquired in the lifetime of a parent could be passed to its offspring.

In fact, very little of this story is true. The truth is Lamarck was more of a philosopher than a scientist. And his book was more of a layman's description of current evolutionary thinking designed for a general audience than a treatise of scientific analysis. Lamarck *did* promote the concept of "inherited acquired traits," but he also promoted the concept of evolution—and he didn't come up with either one, nor did he pretend to. At the time, the notion of inherited acquired characteristics was widely held, including by Darwin. Darwin even praised Lamarck in *Origin of the Species* for helping to popularize the idea of evolution.

Unfortunately for him, poor Jean-Baptiste became the victim of a schoolbook version of the theory he didn't develop. Somewhere along the line a science writer (whose name is lost to history) *acquired* the notion that Lamarck was responsible for the idea of inherited acquired traits, and generations of successive science writers have *inherited* that idea and passed it on. In other words, somebody blamed the theory on Lamarck, and lots of other people have repeated it, right up to today. Textbooks still tell of silly-sounding Lamarckian researchers attempting to prove their theories by cutting off the tails of generation after generation of mice, waiting in vain for a generation to be born without tails.

Here's the funny thing—the theory of inherited acquired traits that's responsible for Lamarck's general disregard? It isn't exactly right, but it may not be exactly wrong.

———

LET'S LEAVE THE story of the fellow guilty of nothing more than repeating the widely accepted theories of his time and turn to a woman who offered theories widely dismissed in hers. Barbara McClintock was the Emily Dickinson of genetics—a brilliant, influential, revolutionary thinker who was ignored by her peers for most of her life. She received her Ph.D. in 1927, when she was twenty-five years old. For the next fifty years, she pursued her singular ideas with little need for—and little receipt of—recognition or encouragement.

Most of her research focused on the genetics of corn—its DNA, its mutation, and its evolution. As I've said, just about every geneticist in the twentieth century believed that genetic mutations were random, rare, and relatively small. But in the 1950s, McClintock produced evidence that in certain circumstances, parts of the genome actively triggered much larger changes. This wasn't evidence of minor mutations in which a slight change in one gene on one chromosome slipped through the proofreading system; this was evidence of seismic changes on the genetic scale. Especially when the plants were stressed, McClintock discovered whole sequences of DNA moving from one place to another, even inserting themselves into active genes. When these genes cut and pasted themselves from one place in the corn's DNA to another, they actually affected nearby genes—by changing the sequence of DNA, they sometimes turned genes on and sometimes turned them off. What's more, McClintock found that these wandering genes weren't behaving completely randomly—there was a method to their meandering. First of all, they relocated to certain parts of the genome more often than to other parts. Second, these active mutations appeared to be triggered by outside influences, by changes in the environment that threatened the survival of the corn, like extreme heat or drought. In short, the corn plant seemed to be engaged in some sort of intentional mutation—neither random, nor rare.

Today, the genetic nomads McClintock discovered are called "jumping genes," and they have reshaped our understanding of mutation and evolution. But widespread acceptance of her thinking was a long time coming. When she first presented her ideas in 1951 at the famed Cold Spring Harbor Laboratory on Long Island, where she worked, she might as well have been jumping up and down for all the respect she received. Instead of being toasted, she was greeted with that tired brew of skepticism and scorn that all too often welcomes fresh thinking of any kind.

Over the next thirty years, as molecular biology and genetics evolved themselves, others slowly began to appreciate McClintock's work. Jumping genes were found in other genomes, beyond corn. Our understanding of mutation began to shift.

In 1983, at the age of eighty-one, Barbara McClintock received a Nobel Prize. With characteristic focus, she continued to look past current thinking and, in her acceptance speech, imagined a future when

> attention undoubtedly will be centered on the genome, with greater appreciation of its significance as a highly sensitive organ of the cell that monitors genomic activities and corrects common errors, senses unusual and unexpected events, and responds to them, often by restructuring the genome.

McClintock's discovery of the "jumping gene" opened the door to the possibility of much more robust mutations than the random and rare that theory allowed. This, in turn, suggested that evolution itself could be faster and more sudden than ever before imagined. Instead of a minor spelling error in one word in one verse of the DNA songbook, whole melody lines could insert themselves all over the genome. Like a good hip-hop artist, the genome has the ability to "sample" itself, creating different, but similar, riffs. And a sturdy, networked genome—the emerging notion of a genome that could cope with problems like an active gene's being knocked out—could often survive, and sometimes benefit, from such improvisation.

Scientists are still only beginning to understand how jumping genes—or transposons, as they're known—actually work. Sometimes they copy and paste—copying themselves and then inserting the new material elsewhere in the genome while remaining in their original location. Other times they cut and paste—removing themselves from their starting place and inserting themselves somewhere else. Sometimes the new genetic element stays in place, and sometimes it's removed by the proofreading system or suppressed by other methods.

This much is clear—sometimes, these transposable genetic elements remain in an active gene once they've inserted themselves, and they make a difference. A recent study demonstrated just how much difference a jumping gene can make under the right conditions. A jumping gene in one line of fruit flies turned the line into semi-superhero fruit flies (researchers aptly named the fly "Methuselah"), with the ability to resist starvation and withstand high temperature, as well as a life expectancy that was 35 percent longer than usual.

The key question for scientists to unravel now is *why* these transposons get the urge to jump. McClintock believed that the jumps are a genomic response to internal or environmental stress that cells can't handle under their existing setup. Essentially, a challenge to survival triggers the organism to throw the mutation dice, hoping it will land on a change that will help. That's what she thought was going on with the corn plants she was studying—too much heat or too little water triggered the corn to gamble its survival on finding a mutation that could help it survive. When that happens, the proofreading mechanism is suppressed and mutations are allowed to blossom. Then natural selection kicks in to select the adaptive mutations over the maladaptive mutations in future generations and presto, evolution!

McClintock not only observed that jumping genes were jumpiest during times of stress, she also noted that they tended to jump to certain genes more than others. She believed this was intentional—if the jumps were random, they would land with similar frequency across the genome. Instead, she believed the genome directed its jumpers toward those places in the genome where mutations were most likely to have a beneficial effect. In other words, the dice were loaded for the corn's benefit—even if just a little bit.

The extent to which these jumping genes have fascinated scientists is evident in the names they have been given: *gypsy*, *mtanga* (Swahili for wanderer), *Castaway*, *Evelknievel*, and *mariner*. Those aren't genes from any particular species and we're still learning about their various functions, but when most genes are given sexy names like *ApoE4*, it's clear that many scientists are fans of these genes, and entranced by what they can teach us. There's even one called "Jordan" named by Washington University researchers after Michael Jordan's amazing leaping ability.

Today, scientists continue to follow McClintock's lead away from the notion that the genome is a rigid set of plans and that mutation—and thus, evolution—is only triggered by rare and random errors. As Dr. Gregory Dimijian of the University of Texas writes:

> The genome has long been thought of as an archival blueprint of life, a relatively permanent record. Mobile genetic elements [such as McClintock's jumping genes] are replacing that view with one of an ephemeral environment, undergoing continuous remodeling.

In other words, the genome likes to move the furniture around.

A SERIES OF studies in the 1980s and 1990s provided additional insight into the genome's ability to gamble on mutation. The first was documented in an incendiary 1987 report by Harvard researcher John Cairns in the journal *Nature* that used language harkening back to the theory of inherited acquired traits—the theory wrongly assigned to Lamarck. Cairns conducted studies with *Eschericia coli*, a bacteria known to its friends and human hosts as *E. coli*. (And despite the fearful reputation it has earned because bad strains sometimes turn up in the wrong place killing people, *E. coli* does far more good than harm—it's one of the essential bacteria toiling away in your digestive system right now that we discussed earlier.)

E. coli is a digestive workhorse in humans and can come in many different "flavors" or variants, one of which can't naturally digest lactose, a sugar derived from milk. Nothing is a bigger threat—or evolutionary pressure—to bacteria than starvation. So Cairns deprived milk-shunning *E. coli* of any food except lactose. Much more rapidly than chance should have allowed, bacteria developed mutations that allowed them to lose their lactose intolerance. Just as McClintock maintained about her corn plants, Cairns also reported that bacteria appeared to target specific areas of their genome—areas where mutations were most likely to be advantageous. Cairns concluded that the bacteria were "choosing" which mutations to go after and then passing on their acquired ability to digest lactose to successive generations of bacteria. In a statement that amounted to evolutionary heresy, he wrote that *E. coli* "can choose which mutation they should produce" and may "have a mechanism for the inheritance of acquired characteristics." He straight-out raised the possibility of inherited acquired traits; he basically used those words. It was like shouting, "Go Sox" at Yankee Stadium during the ninth inning of the seventh game of the playoffs—with Boston leading by a run.

Since then, researchers have plunged into their petri dishes in attempts to prove, disprove, or just explain Cairns's work. A year after Cairns's report came out, Barry Hall, a scientist at the University of Rochester, suggested that the bacteria's ability to happen upon a lactose-processing adaptation rapidly was caused by a massive increase in the mutation rate. Hall called this "hypermutation"—sort of like mutation on steroids—and, according to him, it helped the bacteria to produce the mutations they needed to survive about 100 million times faster than the mutations otherwise would have been produced.

In 1997, other studies added credibility to the hypermutation theory. A significant increase in mutation rates was noticed when *E. coli* were starved of their normal diet but surrounded by lactose. These studies reported an uptick in mutation across the bacterial genome—many different mutations, not just the targeted mutations designed to overcome lactose intolerance that Cairns observed. But even though these researchers reported a greater range of mutation than Cairns documented, the overall increase in mutation also suggests that the genome has the ability to order mutations on demand when the regular genetic programming just isn't good enough. And French researchers led by Ivan Matic, of the Institut National de la Santé et de la Recherche Médicale, studied hundreds of bacteria from all over the world and found that they also went into hyperdrive, mutationally speaking, when put under stress. Although the evidence is mounting, the case of hypermutation is definitely still pending.

CRAZY CORN, A gene named after an NBA basketball player, and lactose-intolerant bacteria are all well and good—but you're probably wondering what all this has to do with us. Before we dive into the human gene pool, let's review a few rules, starting with a generally accepted genetic principle called the Weismann barrier. August Weismann was a nineteenth-century biologist who developed the germ plasma theory, which divides the body's cells into two groups, germ cells and somatic cells. Germ cells are cells that contain information that is passed on to your children. Eggs and sperm are the ultimate germ cells. Every other cell in your body is a somatic cell—red blood cells, white blood cells, skin cells, hair cells are all somatic cells.

The Weismann barrier stands between germ cells and somatic cells: the theory maintains that information in somatic cells is never passed on to germ cells. So a mutation that occurs on the somatic side of the barrier, say, in a red blood cell can't move over to the germ side and, thus, will never be passed on to your children. That doesn't mean a mutation in the germ line can't affect somatic cells in your offspring. Remember that all of the instructions to build and maintain your body originated in the germ line of your parents. So a mutation in your germ line that changes the instructions for hair color would affect the hair color of your children.

The Weismann barrier is an important organizing principle in genetic research, but some research suggests that it isn't as impenetrable as we once thought. Some retroviruses or viruses, which we'll discuss in more detail shortly, may be able to penetrate the Weismann barrier and carry DNA from somatic cells to germ cells. If so, that would theoretically open the door to the idea that acquired adaptations could be passed on to future generations.

Which would mean that Lamarck—discredited for spreading one of many ideas that wasn't his own—got a *really* raw deal.

FROM AN EVOLUTIONARY perspective, we're mostly familiar with germ line mutations—mutations that result in a different gene in egg or sperm that produces a new trait in the offspring. And as you know, when new traits increase the offspring's ability to survive or reproduce, it's more likely to spread throughout the population as the first generation of offspring with the new trait passes it on to the next. When a new trait inhibits survival or reproduction, it will ultimately disappear, as those who carry it are less likely ultimately to survive. But mutations occur outside the germ line all the time. Cancer, of course, is one of the most common—and one of the most frightening—examples. At its most basic, cancer is uncontrolled cell growth caused by a mutation in the gene that is supposed to control the growth of the cancerous cells. Some cancers are at least partially hereditary—mutations in the *BRCA1* or *BRCA2* genes significantly increase the risk of breast cancer, for example, and those mutations can be passed from one generation to another. Other cancers can be caused by mutations that are caused by external triggers—like smoking or exposure to radiation.

It's true that most mutations—especially somatic mutations, like the mutations in lung cells that can be caused by smoking— don't work out so well. That makes sense. Biological organisms, especially humans, are pretty complicated. But mutation, by definition, isn't *necessarily* bad; it's just different. And that, it turns out, may be the key to how jumping genes help humans in two very important ways.

Jumping genes are very active in the early stages of brain development, inserting genetic material all over the developing brain, almost helter-skelter, as a normal part of brain development. Every time one of those jumpers inserts or changes genetic material in brain cells, it's technically a mutation. And all of that genetic jumping around may have a very important purpose—it may help to create the variety and individuality that make every brain unique. This developmental frenzy of genetic copy and paste only happens in the brain, because that's where we benefit from individuality. But as the lead author of the study that discovered this phenomenon, Professor Fred Gage said, "You wouldn't want that added element of individuality in your heart."

The neural network in your brain isn't the only complex system that welcomes diversity—your immune system does too. In fact, your immune system employs what has got to be the most diverse workforce in history; we wouldn't have survived long as a species without it. In order to fight the huge array of potential microbial invaders that threaten us, the human immune system employs more than a million different antibodies—specialized proteins that target specific invaders. The mechanism through which we produce all those different proteins isn't completely understood, especially because we don't have nearly enough genes to explain it (remember, there are only about 25,000 active, coding genes, and we're talking about the possibility of more than a million different antibodies). But new research led by scientists from Johns Hopkins has linked the immune system's antibody production mechanism to the behavior of jumping genes.

B-cells are the basic building blocks for antibodies. When we need to produce a specific antibody, B-cells seek out the instructions for that antibody in their DNA, although the individual lines of instruction are usually mixed in with instructions for other antibodies. They snip away the lines of instruction for other antibodies and sew the rest back together, essentially rewriting their own genetic code and producing a specialized product in the process. This is called V(D)J recombination, named after the regions where the genes that are used in this seek-snip-and-sew trick are found.

This process sounds similar to the cut-and-paste mechanism employed by some jumping genes, but there is one key difference—instead of a neat connection, V(D)J recombination leaves a little loop when it reconnects the remaining strands. Scientists had never seen this loop effect in jumping genes, until the Johns Hopkins team found it in the common fly where a jumping gene called *Hermes* behaves just like V(D)J. Nancy Craig, one of the scientists behind the study, said:

Hermes behaves more like the process used by the immune system to recognize a million different proteins . . . than any previously studied jumping gene. It provides the first real evidence that the genetic processes behind . . . [antibody] diversity might have evolved from the activity of a jumping gene, likely a close relative of Hermes.

Once your body develops antibodies against a specific invader, you always have those antibodies—which often give you a leg up if that invader tries again. Sometimes, that even makes you immune to future infections, like most people are after having had measles. But while the mutations that happen in our B-cells are ours to keep, we can't pass them on to our children—they're on the somatic side of the Weismann barrier. Babies are born with a very small number of antibodies, and their immune systems have to start in overdrive. That's one of the many reasons breast-feeding is good for babies—breast milk contains some of the mother's antibodies, which act as a temporary passive vaccination against infections until the baby's immune system is up and running. We're only just beginning to understand the role that transposable elements—jumping genes—play in life and evolution. They clearly play a much bigger role than we've understood to date. Fully one-quarter of active—coding—human genes show evidence that they've incorporated DNA from jumping genes.

Jef Boeke, a professor of molecular biology and genetics at the Johns Hopkins School of Medicine, suggests that jumping genes

have been remodeling host genomes more than previously realized. . . . These changes were probably frequently disastrous, but occasionally they might have benignly increased genetic variation or even improved survivability or adaptability. Such remodeling probably happened thousands of times during human evolution.

We now know that there have been periods of such massive environmental shift it's hard to imagine random, incremental changes providing enough adaptation to let us survive. Prominent evolutionary thinkers Stephen J. Gould and Nils Eldredge advanced the theory of punctuated equilibrium—the notion that evolution was characterized by a state of general equilibrium punctuated by periods of significant change that were brought about by large environmental shifts. Is it possible that jumping genes helped species adapt their way through those evolutionary exclamation points? You bet.

Jumping genes are beginning to look like Mother Nature's version of on-the-fly genetic engineering. The more we understand how they work, the more they may reveal about how our immune systems protect us against disease and how our very genetic structure responds to environmental stress. This could open up whole new avenues to immunize people against disease, restore compromised immune systems, and even reverse dangerous mutations on a genetic level.

REMEMBER ALL THAT "junk DNA"? That's the stuff that we now call noncoding DNA because it doesn't contain the genetic code to build any cells directly. If you're wondering why we would give millions of strands of DNA a piggyback through evolution, you're not alone. That's why scientists called it junk in the first place. But scientists have now begun to decipher the mystery of those noncoding genes. And it was jumping genes that first provided a key.

Once the scientific community recognized that jumping genes were real—and important—researchers started to look for them in genomes of all kinds, including humans. Their first surprise was that a large portion of our noncoding DNA is made up of jumping genes—as much as half of it. But the bigger surprise was this—those jumping genes look an awful lot like a very special type of virus. You heard that right—a huge percentage of human DNA is related to viruses.

You may think about viruses every day—at least about how to avoid them, whether it's the computer or the biological variety—but it's probably been a while since you read about one in a biology book, so here's a quick refresher. A virus is a snippet of genetic instructions that cannot reproduce on its own. Viruses can only reproduce by infecting a host and then hijacking the host's own cellular machinery. They may replicate themselves thousands of times inside a cell before eventually bursting its walls and moving into new cells. Most scientists don't consider viruses to be "alive," because they can't reproduce or metabolize on their own.

Retroviruses are a very special subset of viruses. In order to understand what makes them so important, it helps to understand how genetic information is used to build cells and, ultimately, organisms. Generally speaking, body building follows this pathway—DNA to RNA to protein. Think of DNA as a library of master blueprints for a whole town and all the different cells in your body as different kinds of buildings—schools, municipal buildings, houses, apartment buildings. When an organism needs to build a particular building, it uses a helper enzyme called RNA polymerase to copy the plans for that building onto strands of messenger RNA, or mRNA. The mRNA takes those instructions to the building site and directs construction of whatever building—or protein—is called for.

For a long time, scientists thought genetic information flowed in only that one direction, DNA to RNA to protein. The discovery of retroviruses—like HIV—proved that wrong. Retroviruses are made of RNA. Using an enzyme called reverse transcriptase, they transcribe themselves from RNA into DNA—they actually reverse the information flow. It's sort of like the messenger rewriting the master blueprint instead of copying and carrying the plans. The implications of this are huge; retroviruses can literally change your DNA. The discovery of RNA that could backslide into DNA led to the development of the novel drugs that are the current mainstay in the "cocktail" therapy used to treat HIV infection. Like a wheel block used by truckers to park their loads, some of these drugs stop the reverse transcriptase enzyme in its tracks: that leaves HIV stuck within the nuclear truck stop, trying to hitch a ride on DNA but unable to climb on board.

Now imagine what happens when a retrovirus or virus writes itself into the DNA of cells in the germ line of an organism. That organism's offspring is born with the virus permanently encoded in its DNA. (By the way, scientists don't think that HIV breaks through the Weismann barrier and inserts itself into the DNA of eggs or sperm. Instead, they believe infected mothers pass HIV to their babies during birth when there's a significant opportunity for the mother's blood to mingle with the infant's.)

Usually, of course, as with all mutations, when an organism's offspring is born with DNA that has been changed by a retrovirus in one of its parents' germ cells, that change is probably harmful, so it doesn't last. But if the virus doesn't hurt—or even helps—the offspring's chance to survive and reproduce, that virus may end up a permanent part of the gene pool. If genetic code that originally came from a virus is part of an organism's gene pool, it's pretty hard to say where one ends and the other begins—virus and organism have become one and the same. Today, we know that at least 8 percent of the human genome is composed of retroviruses and related elements that have found a permanent place in our DNA—they're called HERVs, or human endogenous retroviruses. Scientists are only beginning to uncover the role HERVs play in human health, but they've already found interesting connections. One study showed that a particular HERV may play an important role in the construction of a healthy placenta; another documented links between HERVs and the skin disorder psoriasis.

And those frisky jumping genes? They may very well be descended from viruses, too. There are two basic types of jumping genes—the first type are called DNA transposons and they jump through a cut-and-paste process; the second type, retrotransposons, are copy-and-paste jumpers. It turns out that copy-and-paste jumping genes—*retro*transposons—look an awful lot like retroviruses. That makes sense, because the mechanism those copy-and-paste genes use to insert themselves in other genes is very similar to the mechanism retroviruses use. First, a retrotransposon copies itself onto RNA like any normal gene. Then, when the RNA reaches the place in the genome the jumper wants to land in, the retrotransposon uses reverse transcriptase to paste itself into the DNA, reversing the normal information flow just like a retrovirus does.

Does that mean retro jumping genes are descended from retroviruses?

NOBODY BELIEVES IN the power of viral marketing like Luis Villarreal does. At least, nobody believes there's anything on earth that's better than a virus at spreading its message, getting into everything, and generally outlasting the competition. Villarreal is the director of the Center for Virus Research at the University of California at Irvine, and he's followed the implications of viral impact on human evolution to the limit.

Villarreal gives Salvador Luria, a Nobel Prize–winning microbiologist whose work stretched from the 1940s to the 1980s, credit for the first suggestion that viruses have helped to spark human evolution from the inside, not just the outside. In 1959, Luria wrote that the movement of viruses into genomes had the potential to create "the successful genetic patterns that underlie all living cells."

Villarreal speculated that this idea didn't catch on very quickly because people react with a kind of visceral disgust to the suggestion that we've been shaped by parasites:

> There's a very strong cultural, negative response to the concept of a parasite of any kind. The irony is that ... this is such a crucial creative force. . . . If you want to evolve, you have to be open to being parasitized.

In his book *Viruses and the Evolution of Life*, published in 2005, Villarreal argues it's high time to take a fresh look at viruses. Villarreal distinguishes familiar, deadly parasites like HIV and smallpox from those he calls "persisting viruses." Persisting viruses are the viruses that have migrated into our genome over millions of years and may have become our partners in evolution.

It seems clear what the viruses get out of a permanent home in our genomic mother ship—a free ride through life. But what do we get out of it? Well, viruses are master mutators—they are vast storehouses of genetic possibility and they can deliver that possibility incredibly fast, mutating as much as a million times faster than we do. To drive home the sheer volume of genetic potential in the viral world, Villarreal often asks people to try to imagine all of the viruses in the world's oceans—all 100,000,000,000,000,000, 000,000,000,000,000 of them (that's 100 nonillion for those of you who are counting). These little containers of genetic code are microscopic, but if you laid them end to end they would be 10 million light-years long. By tomorrow, most of them will have spawned a new generation—and that's what they've been doing for several billion years. Villarreal calls viruses "the ultimate genetic creators, inventing new genes in large numbers, some of which find their way into host lineages following stable viral colonization."

Here's how that works for us. Persisting viruses in our genome have as much at stake in our survival and reproduction as we do—since they're part of our DNA, they've got an evolutionary interest in our success. Over the last few millions of years, perhaps we've given them the ride of their life and, in return, they've given us the chance to borrow some code from their huge genetic library. With all that mutating power, they are bound to happen on useful genes far faster than we could without their help. Essentially, this partnership with viruses may have helped us evolve into complex organisms much faster than we would have on our own.

The study of jumping genes has produced evidence that bolsters Villarreal's theory. As we've discussed, jumping genes are probably descended from viruses. As it turns out, the more complex an organism is, the more jumping genes it has. Humans and our African primate relatives even share a particular genetic trait that makes it easier for our genomes to do business in the viral marketplace. Our genomes have been modified by one particular retrovirus in a way that makes it easier for us to be infected by other retroviruses. According to Villarreal, this capacity of African primates to support the persistent infection of other viruses may have put our evolution on "fast forward" by allowing more rapid mutation through exposure to other retroviruses. It's possible that this capacity helped spur our evolution into humans.

Which means that all that "junk DNA" may have possibly provided the code for our evolution up and away from our furry cousins. Which means that viruses may have infected us with that code. Which means—

Infectious design, anybody?

CHAPTER VII

METHYL MADNESS:
ROAD TO THE FINAL PHENOTYPE

One-third of American children are overweight or obese—that's 25 million kids. In the last thirty years, the percentage of obese two- to five-year-olds has doubled—and the percentage of obese six- to eleven-year-olds has tripled. A baby girl born in 2000 now has a *40 percent chance*—almost a coin toss—of developing of Type 2 diabetes, and that's directly related to the huge surge in heavy kids.

What's even sadder is that many of these children are showing symptoms of obesity-related illness while they're still kids. One recent study showed that about 60 percent of obese five- to ten-year-olds already exhibited at least one major risk factor for heart disease—high cholesterol, high blood pressure, high triglycerides, or high sugar levels. Of those kids, 25 percent had more than one risk factor. A 2005 report in *The New England Journal of Medicine* said that the epidemic of childhood obesity is the critical element in a gathering storm that could produce the first modern *decline* in American life expectancy—dropping life expectancy as much as five years.

There's no question that gallons of sugary soda, baskets of fatty fries, and too many hours watching television and playing video games instead of after-school sports is a fattening combo. But new research suggests that may not be the whole story.

There is emerging evidence that the dietary habits of parents, especially women in the earliest stages of pregnancy, may have an impact on the metabolism of their children. In other words, if you're trying to get pregnant, you really should think twice before you bite that Big Mac—once for your own waistline, and once for your potential child's.

Before you get the wrong idea, this isn't to suggest some strictly Larmarckian idea that a fat parent is going to have a fat child because the child will *inherit* the weight problem his or her parent *acquired*. But this *is* to say that new research is rapidly changing our understanding of how, when, and whether genes express themselves—that is, how, when, and whether the instructions in a gene are carried out. A series of groundbreaking research over the last five years has shown that certain compounds can attach themselves to specific genes and suppress their expression. These compounds act like a genetic light switch, essentially turning off the genes they attach to. And—here's where it gets really interesting—the research shows that environmental factors, like the food we eat or the cigarettes we smoke, can flick the switch on or off.

This research is changing the whole field of genetics—it's even launched a subdiscipline called epigenetics. Epigenetics is concerned with the study of how children can inherit and express seemingly new traits from their parents *without* changes in the underlying DNA. In other words, the instructions are the same, but something else overrides them.

Being a gene isn't all that it was cracked up to be anymore.

THE TERM *EPIGENETICS* was coined in the 1940s, but the modern discipline is much younger, barely out of diapers. The first big breakthrough actually occurred in 2003—in the form of a skinny brown mouse.

The shocking thing about this skinny brown mouse is that its parents were both fat yellow mice. Actually, they were fat yellow mice from a long line of fat yellow mice. These mice were specifically bred to carry a gene called *agouti*, which gives them their characteristic pale coat and tendency toward obesity. When a male *agouti* mouse mates with a female *agouti* mouse, they have little *agouti* mouse babies time after time—fat and yellow. Or they did until they went to Duke, anyway.

A team of scientists at Duke University separated a gang of *agouti* mice into two groups—a control group and a pregnant group. They didn't do anything special with the control group. They fed it a normal diet and let fat yellow Mickeys mate with fat yellow Minnies, who gave birth to fat yellow babies. No surprise there.

The mice in the experimental group mated as well, but the expectant mothers in this group got slightly better prenatal care—in addition to their normal diet, they were given vitamin supplements. In fact, they were given a combination of compounds that is a variation on the prenatal vitamins given to pregnant women today—vitamin B_{12}, folic acid, betaine, and choline.

The results rocked the genetic world. Fat yellow female mice that had mated with fat yellow male mice had thin brown babies. That seemed to throw everything the scientific community understood about heredity up in the air. A genetic examination of the brown baby mice only added to the mystery. Their genes were the same as their parents'. The *agouti* gene in the thin brown mice was right where it was supposed to be, ready to send out instructions to make them fat and yellow. So what happened?

Essentially, one or more of the compounds in the vitamin supplements fed to the expectant mothers reached down into the mouse embryos and flicked the *agouti* gene into the "off" position. When the baby mice were born, their DNA still contained the *agouti* gene, but it wasn't expressed—chemicals had attached to the gene and suppressed its instructions.

This process of genetic suppression is called DNA methylation. Methylation occurs when a compound called a methyl group binds to a gene and changes the way that gene expresses itself, without actually changing the DNA. The compounds in the vitamin supplements include methyl donors—molecules that form the methyl groups that become these genetic stop signs.

Thin and brown weren't the only benefits the mice gained through methylation. The *agouti* gene in mice is linked to higher rates of diabetes and cancer. The mice with the switched-off *agouti* genes had significantly lower rates of cancer and diabetes than their parents.

Of course, we've long understood the basic idea that good nutrition in an expectant mother is important for infant health. And we've also known that the connection goes beyond the obvious—sufficient nutrition, healthy birth weight, and so forth—to reduce the likelihood of certain diseases later in life. But until the Duke study, the "how" was very unclear. As Dr. Randy Jirtle, one of the leaders of the study, said:

> We have long known that maternal nutrition profoundly impacts disease susceptibility in their offspring, but we never understood the cause-and-effect link. For the first time ever, we have shown precisely how nutritional supplementation to the mother can permanently alter gene expression in her offspring without altering the genes themselves.

The impact of the Duke study was enormous, and the study of epigenetics has exploded since it was published. You can imagine why.

First, epigenetics erased the conviction that genetic blueprints are written in indelible ink. Suddenly, science had to take into account the notion that a given set of genes is *not* an immutable set of blueprints or instructions. The exact same set of genes can produce different outcomes depending on which genes have undergone methylation and which have not. There was a whole new layer to consider—a set of reactions that acted outside and above the genetic code, changing its result without changing the code itself. (That outside and above is where epigenetics gets its name—from the Greek prefix *epi,* meaning upon, after, or in addition.) This shouldn't have been a *complete* surprise—for fifty years, some researchers have pointed out that the same genes don't always produce the same results: identical twins (who share identical DNA) don't get the same diseases or fingerprints, just similar ones.

Second, the Duke study snuggled right up to the ghost of Lamarck. Environmental factors in the life of the mother were shown to affect the inheritance of traits in her offspring. These factors didn't change the DNA the baby mice inherited, but in changing the way the DNA was expressed, they changed heredity.

After those first mice experiments, other scientists at Duke showed that they could supercharge the brains of mice simply by adding a touch of choline to a pregnant mouse's diet. The choline triggered a methylation pattern that turned off the gene that normally acted to limit cell division in the memory center of the brain. With the cell division governor turned off, these mice started producing memory cells in high gear—and sure enough, they developed mighty mouse memories. Their neurons fired more rapidly and could fire more often. As adults, these megabrain mice broke all the records in all the mazes.

RESEARCHERS WHO STUDY all kinds of animals—from mammals to reptiles to insects—have long noted the ability of some organisms to produce offspring that seem to be custom-tailored on the basis of the mother's experiences during pregnancy. They noted this ability—but they couldn't really explain it. Once scientists understood the possibility of epigenetic influence on heredity, it all made a lot more sense.

The vole is a furry little rodent that looks something like a fat mouse. Depending upon the time of year its mother is due to give birth, baby voles are born with either a thick coat or a thin coat. The gene for a thick coat is always there—it's just turned on or off depending on the level of light the mother senses in her environment around the time of conception. The developing genome basically gets a weather forecast before it has to go out into the world, so it knows what kind of coat it should grow.

The mother of the tiny freshwater flea *Daphnia* (which isn't really a flea at all; it's actually a crustacean) will produce offspring with a larger helmet and spines if it's going to give birth in an environment crowded with predators.

The desert locust lives in two remarkably different styles depending on the availability of food sources and the density of the local locust population. When food is scarce, as it usually is in their native desert habitat, locusts are born with coloring designed for camouflage and lead solitary lives. When rare periods of significant rain produce major vegetation growth, everything changes. At first, the locusts continue to be loners, just feasting off the abundant food supply. But as the extra vegetation starts to die off, the locusts find themselves crowded together. Suddenly, baby locusts are born with bright colors and a hankering for company. Instead of avoiding one another and hiding from predators through camouflage and inactivity, these locusts gather in swarms, feed together, and overwhelm their predators through sheer numbers.

One species of lizard is born with a long tail and large body or a small tail and small body depending on one thing only—whether their mother smelled a lizard-eating snake while pregnant. When her babies are entering a snake-filled world, they are born with a long tail and big body, making them less likely to be snake food.

In each of these cases—the vole, the water flea, the locust, and the lizard—the characteristics of offspring are controlled by epigenetic effects that occur during fetal development. The DNA doesn't change—but the way it's expressed does. This phenomenon—the mother's experiences influencing gene expression in her offspring—is called a predictive adaptive response or maternal effect.

IMAGINE THE IMPLICATIONS of this for humans. By sending the right epigenetic signals, we can have healthier, smarter, better-adapted babies. As we learn more, we may be able to suppress the genes that express themselves in harmful ways even after birth—or turn helpful genes back on after they have been turned off. Epigenetics has the potential to give us a whole new measure of control over our health. DNA is destiny—until you get out the old methyl Magic Marker and start rewriting it.

The current focus in human epigenetics is on fetal development. It's now clear that the first few days after conception—when a mother may not even know she's pregnant—are even more critical than we've understood. That's when many important genes are switched on or off. And the earlier that epigenetic signals are transmitted, the more significant the potential changes are in the fetus. (In some ways, the womb may be like a tiny evolutionary laboratory, examining new traits to see whether they'll help the fetus survive and thrive; if they won't, the mother miscarries. Researchers have certainly noted that many miscarried fetuses have genetic abnormalities.)

Here's how epigenetics may be partially responsible for the epidemic of childhood obesity. The junk food that fills so many American diets is high in calories and fats, but often very low in nutrients, especially those that are important to a developing embryo. If a newly pregnant mother spends the first weeks of her pregnancy eating a typical junk-food-laden diet, the embryo may receive signals that it's going to be born into a harsh environment where critical types of food are scarce. Through a combination of epigenetic effects, various genes are turned on and off and the baby is born small, so it needs less food to survive.

But that's only half the story. Almost twenty years ago, a British medical professor named David Barker (who won the Danone International Prize for nutrition in 2005) first suggested a link between poor fetal nutrition and later obesity. His theory, known as the Barker Hypothesis or the thrifty phenotype hypothesis, has been gaining ground ever since. (Phenotype is the physical expression of your genotype; in other words, if you have one parent with attached earlobes and the other parent with detached earlobes, you will have detached earlobes, because that trait is dominant—detached earlobes would be part of your phenotype. Epigenetic effects influence your phenotype without changing your genotype. So, in this hypothetical example, if a methyl marker turned off your gene for detached earlobes, your *phenotype* would change—you'd have attached earlobes—but your genotype would remain the same. You'd still have the gene for detached earlobes to pass on to your children in either the on or off state; it would just be deactivated in you.) According to the thrifty phenotype hypothesis, fetuses that experience poor nutrition develop "thrifty" metabolisms that are much more efficient at hoarding energy. When a baby with a thrifty phenotype was born 10,000 years ago during a time of relative famine, its conservationist metabolism helped it survive. When a baby with a thrifty metabolism is born in the twenty-first century surrounded by abundant food (that is also often nutritionally poor but calorie rich), it gets fat.

Epigenetics makes the thrifty phenotype hypothesis even more compelling, because it helps us to understand how a mother's eating habits could affect the metabolic makeup of her children. If you're thinking about having a baby, you're probably already asking yourself what you should eat and when during your pregnancy. We don't know enough yet to understand exactly when human fetuses reach epigenetic trigger points. But animal studies suggest the process starts very early.

One recent study of rats showed that when pregnant rats were fed a low-protein diet for just the *first four days* of pregnancy— before the embryo had even implanted in the uterus—their babies were prone to high blood pressure. Experiments with sheep showed similar maternal effects. Pregnant sheep that were underfed during the early days of pregnancy—again, even before the embryo implanted in the mother's uterus—gave birth to offspring that rapidly developed thickened arteries because their slower metabolisms stored more food as fat.

How do we know these are adaptive responses, as opposed to birth defects resulting from the mother's poor nutrition? Because the health problems—thickened arteries and increased weight— only occurred when the baby sheep were provided with normal diets. Baby sheep whose mothers were undernourished while pregnant showed no sign of arterial thickening when they were also undernourished as toddlers.

Most of the epigenetic effects currently under study involve mothers, not fathers. In part, that's because an embryo or fetus never interacts with its father's environment, so many scientists believed epigenetic modifications only occurred after conception, in response to information the fetus received about the mother's environment. However, there is new and intriguing evidence that fathers can pass information to their offspring as well. A British study found that men who started to smoke before puberty had sons who were significantly fatter than normal by the time they were nine; this correlation was found only in sons, so scientists think these epigenetic markers are passed on the Y chromosome. (Intuitively, you might expect the children of smoking fathers to be *smaller*, not fatter. It's possible that this effect is analogous to the thrifty phenotype, in which poor maternal nutrition in the early stages of pregnancy leads to the birth of small babies with thrifty metabolisms who have a high tendency to become fat. In this case, there may be an epigenetic change in the father's sperm triggered by the toxins in the smoke the father is inhaling. Those toxins would indicate a difficult environment, so the sperm is ready to create a baby with a thrifty metabolism. And when that thrifty metabolism is combined with a typical Western diet, the likelihood of that baby growing up to be a fat child dramatically increases.)

The lead scientist on the study, Marcus Pembrey, a British geneticist, believes this proves the existence of paternal effects in addition to maternal effects. He called this "proof of principle. The sperm have captured information about the ancestral environment, and this is modifying the development and health of subsequent generations."

This lends a whole new meaning to sons paying for the sins of their fathers.

MOM AND DAD may not be the only epigenetic influences in your life. Grandpa and Grandma may be reaching down from their perch above you in the family tree, leaving their own marks. That's certainly what many of the most prominent epigenetic researchers—from the authors of the fat yellow mice study at Duke to the researchers behind the smoking fathers report in

London—think. They all believe that epigenetic changes can be passed through the germ line for many generations.

In the case of maternal inheritance, the opportunity for your ultimate genotype to get a methyl markup in your grandmother is actually very direct. When a human female is born, she already has the complete set of eggs she will have for life in her baby ovaries. As strange as it sounds, that means that the egg you developed from, with half of your chromosomes, was created in your mother's ovaries while she was still in your grandmother's womb. And new research demonstrates that when your grandmother passed epigenetic signals to your mother, she was also passing those signals to the egg that would eventually provide half of your DNA.

Just as epigenetics has helped to unlock the mystery of thin-coated voles and sociable locusts, it's now helping to explain a series of confusing correlations researchers have gathered over the last century. A group of researchers in Los Angeles found that children whose grandmothers smoked while pregnant were more likely to have asthma than children whose mothers smoked while pregnant. Before we started to crack the epigenetic code, this correlation was impossible to explain. Now, scientists realize that the smoking grandmother triggered an epigenetic effect in her fetal daughter's supply of eggs. (Incidentally, if you're puzzled as to why the grandmothers' smoking habits affected their eggs more than their fetuses, you're not alone; scientists haven't figured that out yet.)

A harsh winter and a cruel embargo imposed by the Nazis combined to cause the Dutch famine of 1944 and 1945. Thirty thousand people died during the "Hunger Winter," or Honger-winter, as the Dutch call it. An examination of birth records following the famine is one of the ways Barker confirmed his thrifty phenotype hypothesis. Women who were in the first six months of pregnancy during the Hongerwinter gave birth to small babies who grew up to be more prone to obesity, coronary disease, and a variety of cancers.

Although the results are still controversial, researchers reported an even bigger surprise around twenty years later when their studies indicated that the *grandchildren* of those women were also born with low birth weights. Is it possible that the methyl markers triggered by poor nutrition during the famine were passed on to the next generation? That's not known yet, but the effects of methylation, it seems, are real.

Many leading epigenetic scholars think epigenetic changes represent evolution's subtle effort to tweak an existing genome, although that's still quite contentious. The scientists at Duke who published the mouse study wrote:

> Our findings show that early nutrition can influence the establishment of epigenetic marks ... [that] affect all tissues, including, presumably, the germ line. Hence, incomplete erasure of nutritionally induced epigenetic alterations ... provides a plausible mechanism by which adaptive evolution may occur in mammals.

In other words, when methyl markers aren't erased, they can be passed on generation after generation, ultimately leading to evolution. Or in *other* other words, traits *acquired* by a parent or grandparent can ultimately be *inherited* by his or her descendants. Lamarck must be turning in his grave. The theory that he didn't come up with is on the verge of becoming all the rage. Marcus Pembrey, the scientist behind the parental smoking study, calls himself a "neo-Lamarckian." And Douglas Ruden, a researcher at the University of Alabama, told a reporter from *The Scientist,* "Epigenetics has always been Lamarckian. I really don't think there's any controversy."

MOST OF THE methyl effects we've talked about so far involve changes that take place before birth. But epigenetic changes occur throughout life, as the placement of methyl markers turns some genes off and the removal of methyl markers turns other genes back on.

In 2004, Michael Meaney, a professor at McGill University in Canada, published a report that caused nearly as big a sensation as the Duke report about yellow and brown mice. Meaney's study showed that the interaction between mothers and their offspring *after* birth provoked the placement of methyl markers that caused significant epigenetic changes.

Meaney studied the behavior of rats that received different levels of attention from their mothers in the first few hours after birth. Pups that were gently licked by their mothers grew into confident rat babies that were relatively relaxed and could handle stressful situations. But rats that were ignored by their mothers grew to be nervous wrecks.

Now, this sounds like an experiment ripe for a nature versus nurture debate, doesn't it? Those on the nature side would argue that rat moms with bad social skills passed on their emotionally troubled genes to rat babies that grew up to have bad social skills, while the well-adjusted rats gave their babies well-adjusted genes. That makes sense as far as it goes—except that Meaney and his colleagues pulled a mate-and-switch. They gave babies from standoffish mothers to loving mothers, and vice versa. Pups that were fawned over grew to be calm regardless of their natural mother's behavior.

Are all you nurture advocates out there smelling victory? If rats that were treated well turned out well regardless of their genetic makeup, then that means their personalities developed in response to their parenting. Score one for Mother Nurture.

Not so fast.

An analysis of the rats' genes showed striking differences in methylation patterns between the two sets of rats. Rat pups that were attentively groomed by their mothers (biological or adopted) showed a *decrease* in methyl markers around the genes involved with brain development. The mothers' gentle attention somehow triggered the removal of methyl markers that would otherwise have blocked or impeded the development of a part of their babies' brains—almost as if they were licking them off. The part of the brain that dampened the stress response was more developed in those babies. This wasn't nature *versus* nurture; this was nature *and* nurture.

Meaney's paper was another epigenetic blockbuster. Something as simple as parental grooming was *changing the expression of a living animal's genetic code.* The notion was so shocking that some people had a hard time accepting it. One reviewer at a prominent journal actually went so far as to write that, despite the researchers' carefully marshaled evidence, he refused to believe it could be true. It just wasn't supposed to happen like that.

But it does.

WE DON'T REALLY know for sure whether parental care for human infants has the same kind of effect on the development of human brains. In one sense, though, it doesn't matter—because we already know that parent-child bonds from birth through early childhood have a profound impact on emotional development. We know that the emotional state of loving, responsive parents gets passed on to their children in a kind of mental methylation— and so does anything that increases a parent's anxiety. Everything from a dissolving marriage to health problems to financial trouble can raise the stress of a new parent and interfere with the child-parent relationship. Children whose parents are overly stressed are more prone to depression and have less self-control. Children whose parents are relaxed and available tend to be happier and healthier.

And while we don't know whether neonatal parenting is actually changing brain development, scientists who study this epigenetic connection in animals believe it's very unlikely that humans don't share it. In fact, the total picture suggests humans should be *more* prone to epigenetic effects in infancy. After all, cognitive development and physical development after birth in humans are much more significant than they are in most other mammals.

LIKE MUTATION, METHYLATION is neither good nor bad on its own—it all depends on what genes are being turned on and what genes are being turned off and for what reason. Good nutrition in pregnant mice led to the *addition* of methyl markers on the *agouti* gene that freed a generation of baby mice from a fat yellow future. Parental grooming in rats provoked the *removal* of methyl markers around genes responsible for brain development. The same thing is true in humans. Some genes are better turned off, and there are other genes that we want on duty 24/7. Methylation also doesn't always just turn a gene completely off. Genes can be partially methylated, and the degree of methylation correlates to how active the gene remains—the less methylation, the more active it is.

One set of genes that we want always on guard are those that suppress tumors and repair DNA. Those genes are the storm troopers and flight surgeons of the anticancer corps. Scientists have identified dozens of these genetic guardians—when they're shut down, cancerous cells have free rein.

A recent article in *Science News* told the story of two identical twins, Elizabeth and Eleanor (not their real names), who were born on November 19, 1939. From the moment the twins were born, they were treated the same because their mother never wanted either girl to feel she was more—or less—favored. Elizabeth said, "We were treated like a unit—more like one person instead of two separate individuals." They moved apart more than forty years ago, in their early twenties, but they're still very similar. From the way they look to the things they care about, it's clear that they're identical twins. With one big exception—seven years ago, Eleanor was diagnosed with breast cancer. Elizabeth has never been.

Identical twins share the same exact DNA—but DNA isn't fate. And one of the reasons is methylation. It's possible that forty-plus years of exposure to a different environment produced a different methylation pattern around Eleanor's genes, a pattern that unfortunately may have led to breast cancer.

In 2005, Manel Esteller of the Spanish National Cancer Center, along with colleagues, issued a report showing that identical twins shared almost identical methylation patterns at birth that diverged as they grew older. And the report indicated that those patterns diverged much more dramatically when the twins lived apart for most of their lives, just as Eleanor and Elizabeth have. Esteller said:

We believe these different epigenetic patterns in twins depend many times on the environment, whether it's exposure to different chemical agents, diets, smoke, or whether people live in a big city or the countryside.

There's more evidence coming in to support the idea that methylation of specific genes is tightly connected to cancer. In Germany, scientists at a company called Epigenomics have reported an overwhelming connection between breast cancer recurrence and the amount of methylation of a gene called *PITX2*. Ninety percent of the women with low methylation of the *PITX2* gene were cancer-free after ten years, while only 65 percent of the women with high methylation were as lucky. Ultimately, this kind of information will help doctors to custom tailor cancer treatments—the more help they can get from the body's natural cancer fighters, the less aggressive they may need to be in terms of chemotherapy and radiation. The data from Epigenomics is already being used to help women who have low methylation of *PITX2* decide if chemotherapy is necessary after their tumor is removed.

Scientists are establishing clear links between methylation of cancer-fighting genes and cancer-causing behavior. Over time, habits like smoking can cause a massive buildup of methyl markers around these genes. Scientists call this hypermethylation. People who smoke exhibit hypermethylation around genes that would otherwise combat lung cancer. Genes that are supposed to fight prostate cancer are hypermethylated in smokers, too.

In part because of the hypermethylating effect of potentially carcinogenic habits, methylation patterns can also be an early warning signal. In India, millions of people are addicted to betel nuts, a peppery seed that stains the teeth and gums red when it's chewed and, like nicotine, is mildly intoxicating, highly addictive, and seriously carcinogenic. Because of betel nut chewing, oral cancer is the most common cancer in Indian men. And because oral cancer often doesn't manifest any symptoms for a long time, it's often fatal—70 percent of the people diagnosed with oral cancer in India eventually die of it. A lifetime of betel nut chewing can lead to hypermethylation of three cancer-fighting genes—one that suppresses tumors, one that repairs DNA, and one that hunts out lone cancer cells and gets them to self-destruct. Reliance Life Sciences, the Indian company that established this link, has developed a test to measure the degree of methylation in these genes. "We'd like to use the degree of methylation at sites near these three genes as a predictive marker to qualitatively say how far a person is from developing oral cancer," said Dr. Dhananjaya Saranath, one of the scientists at Reliance Life Sciences. Ultimately, tests like this could be an enormous tool in measuring cancer risk, leading to much earlier diagnosis and much higher survival rates.

RIGHT NOW EPIGENETICS is in a bit of a the-more-we-know-the-less-we-understand phase. One thing is clear—it seems pretty certain that things we know to be bad for us can end up being bad for our descendants, as epigenetic markers get passed on from generation to generation. So smoking two packs a day and living a Super-Sized life may actually make your children—and even their children—more prone to disease.

But what about using methyl markers to have a positive influence on our kids? Folic acid and B_{12} worked for mice—will it work for humans? If your family's had a bit of a weight problem as far back as you can remember, can a few methyl markers prevent that heritage from weighing your baby down? The truth is, we just don't know—and we don't even know everything we don't know yet.

Here's the first thing we don't know—we don't have anywhere near a complete understanding of which genes are turned off or turned down by which methyl donors. For example, methylation of a gene that influences hair color might lead to a harmless change—but the same process that triggered methylation of the hair color gene may also be suppressing a tumor suppressor. To complicate things further, methyl stop signs often land near transposons—those jumping genes. When that transposon inserts itself somewhere else in the genome, it may carry methyl markers with it where they may attach themselves to another gene, muting its expression or at least turning down the volume.

In fact, the authors of the Duke study were so impressed by the enormous range of potential epigenetic effects that they issued a word of caution to anyone interested in applying the results of their research to humans:

> These findings suggest that dietary supplementation, long presumed to be purely beneficial, may have unintended deleterious influences on the establishment of epigenetic gene regulation in humans.

In other words, we don't really know everything that's going on here, folks.

To be clear, if you're getting ready to have a baby, this isn't to suggest that you throw out the container of vitamins your doctor prescribed. These vitamins have a lot to recommend them—as we mentioned a few chapters ago, folic acid is very important during pregnancy. Study after study has shown that folic acid supplements reduce birth defects that can cause damage to a developing brain or spinal cord. The connection is so strong that the government required grains to be fortified with folic acid much as drinking water is fortified with fluoride. And there's been a corresponding decrease in diseases, such as spina bifida, that are related to folic acid deficiency in pregnant women.

That's a wonderful thing—but it may not be the whole story. Our understanding of epigenetics is so immature we have to be wary about unintended consequences. We just don't know what other genes may be influenced by pumping methyl donors into the food supply, and we probably won't know for years.

When doctors expect a pregnant woman to give birth prematurely, she is often injected with a drug, usually betamethasone, to help speed up the development of her fetus's lungs, dramatically improving its chance of survival. Now, there are signs that children whose mothers received multiple doses of betamethasone have increased levels of hyperactivity and slower than normal overall growth. A recent University of Toronto study demonstrated that these effects may continue for multiple generations. The leader of the study believes the betamethasone causes epigenetic changes in the fetus that are passed on to its own offspring in turn. One doctor who specializes in treating premature babies said the study was "terrifying beyond comprehension."

Vitamins and drugs that cause methylation in addition to fulfilling their primary purpose are just the beginning. Now we're starting to see drugs actually designed to affect methylation patterns. The first of these drugs was approved by the Food and Drug Administration in 2004. Called azacitidine in its generic form, it was hailed as a breakthrough for the treatment of myelodysplastic syndrome, or MDS. MDS is a collection of blood disorders that is very difficult to treat and often leads to potentially deadly leukemia—a new drug for MDS would be a significant advance. Azacitidine inhibits the methylation of certain genes in blood cells, helping to restore proper DNA function and reducing the risk that MDS will develop into leukemia. Azacitidine was met with tremendous excitement at its introduction. Peter Jones, a professor of biochemistry and molecular biology at the University of Southern California, said:

> This is the first approved drug in a new kind of therapy—epigenetic therapy. That gives it tremendous potential importance not just in this disease, but in a host of others as well.

Of course, in a report by Dr. Jones and some colleagues, he also noted:

> It is apparent that we are just at the beginning of understanding the substantial contribution of epigenetics to human disease and there are probably many surprises ahead.

"Many surprises ahead." Well, he was right. Six months after azacitidine was approved, researchers at Johns Hopkins published a report of their investigation into the epigenetic effects of two drugs, one of them a close chemical relative of azacitidine. These drugs were all but spray painting the genome with new methylation patterns, turning off as many genes as they were turning on—hundreds of each.

Don't get me wrong—epigenetics has unbelievable potential to have a positive impact on human health. A Rutgers University professor named Ming Zhu Fang has studied the effect of green tea on human cell lines. He's found that compounds in green tea inhibit the placement of methyl markers on genes that help to fight colon, prostate, and esophageal cancer. Methylation of those genes would take them out of the cancer suppression business—by inhibiting their methylation, green tea keeps them in the anticancer fight.

The same Duke team responsible for the original study of vitamin-triggered methylation in *agouti* mice has demonstrated a similar methylating effect from genistein, the estrogenlike compound found in soy. They've speculated that genistein may also help to reduce the risk of obesity in humans, perhaps even helping to explain why Asian rates of obesity are comparatively low. But again, their speculation is tempered with a note of caution. Dana Dolinoy, one of the study's authors, said:

> What is good in small amounts could be harmful in large amounts. We simply don't know the effects of literally hundreds of compounds that we intentionally or inadvertently ingest or encounter each day.

There are 3 billion base pairs of nucleotides in the human genome engaged in a vast and complex dance that makes us who we are. We need to be awfully careful when we start to change the choreography, especially given our current lack of precision. When you try to move one dancer with a bulldozer, you're pretty darn certain to scoop up more than one Rockette.

IF THAT'S NOT complicated enough, methyl markers aren't the only way genes are turned on or off. There is a whole system of promoters and repressors that govern how much a given gene expresses itself by transcribing into mRNA and then translating into a protein. This system amounts to an internal regulator that can turn on, turn off, or even crank up production of specific proteins in response to the body's changing needs.

This is how people build up their tolerance to drugs and alcohol, for example. When someone drinks alcohol, the genetic promoters in his or her liver cells crank up production of the enzyme (remember alcohol dehydrogenase?) that helps to break it down. The more you drink, the more your liver produces alcohol dehydrogenase—its biological anticipation of the next drink. And the reverse is also true—you might notice your tolerance drop after a period of sustained teetotaling, because your body slows down the production of alcohol dehydrogenase when it no longer senses the regular need for it.

There's a similar phenomenon with other drugs, from caffeine to many prescription drugs. Have you ever been prescribed a drug that gave you some unpleasant side effects only to have your doctor tell you just to wait a few weeks and they'll go away? If you have, and they've gone away, you've experienced another form of gene expression. Your body adapted to the presence of the drug by promoting or suppressing the expression of specific genes that helped you to process it.

IF YOU REALLY want to understand how *little* we understand about possible epigenetic and maternal effects, consider the following. In the months immediately after the terrorist attacks on New York and Washington on September 11, there was a dramatic spike in the number of late-term miscarriages—in California. It would be tempting to assume that there is an obvious, behavior-related explanation for this—higher stress made it harder for some expectant mothers to take care of themselves. It is tempting to accept this except for one thing—the rise in miscarriages only affected male fetuses.

In California, in October and November 2001, there was a 25 percent increase in the rate of male miscarriages. Something—and we don't know what—in the mother's epigenetic or genetic architecture sensed that she was carrying a boy and triggered a miscarriage.

We can speculate why this occurred, but we really don't know the truth. Males are both more demanding physiologically on the mother's body during pregnancy and less likely to survive if malnourished as children. Perhaps we have evolved a kind of automatic resource conservation system that is triggered in times of crisis—lots of females and a few strong males gives a population a better chance for survival than the other way around.

Whatever the evolutionary reason, it is clear that these pregnant women responded to a perceived environmental threat with a dramatic—and automatic—reaction. The fact that the actual attack occurred so far away only makes it more interesting. And this isn't the first time such a reaction has been documented. During the reunification of Germany in 1990, the birth rate in the former East Germany (where reunification was difficult, tumultuous, and anxiety-producing) skewed toward females. A study of births after the ten-day war in Slovenia during the Balkan conflicts of the 1990s and another study of births after the Hanshin earthquake of 1995 in Kobe, Japan, showed evidence of a similar pattern.

On the other side of the coin, there is evidence that in times after great conflict, the male birth rate goes up. That's what happened after World War I and World War II. A more recent study of six hundred mothers living in Gloucestershire, England, revealed that those who predicted that they would live well into old age were more likely to have male babies than those who predicted that they would die relatively young.

Somehow, an expectant mother's mental state can trigger physiological or epigenetic events that can affect her pregnancy and the relative viability of male or female fetuses. Good times mean more boys. Tough times mean more girls. And epigenetics means we've got more—much more—to learn.

THE FIRST BIG epigenetic breakthroughs were published just as other scientists were announcing the completion of the Human Genome Project—the mammoth ten-year effort to map out the sequence of all 3 billion nucleotide pairs that make up our DNA. When they were done, project organizers announced that they had effectively created "all the pages of a manual needed to make the human body."

And then epigenetics really rained on their parade. After ten years of painstaking work, the scientists came out of their labs to find out that their map was only a starting point. The scientific community might as well have said, "Thanks for the map. Now can you tell us which roads are open and which roads are closed so we can make some use of it?"

Of course, epigenetics doesn't really make the Human Genome Project worthless—to the contrary, a map of the epigenome has to begin with a map of the genome. And sure enough, work has begun to make one. In the fall of 2003, a group of European scientists announced the Human Epigenome Project. Their goal is to add an indicator to every spot where methyl markers can attach and change the expression of a given gene. As they say:

> The goal of the Human Epigenome Project is to identify all the chemical changes and relationships ... that provide function to the DNA code, which will allow a fuller understanding of normal development, aging, abnormal gene control in cancer and other diseases, as well as the role of the environment on human health.

The money is slowly coming in, and they hope to have most of the epigenome mapped in the next few years, but it won't be easy. Science never is.

CHAPTER VIII

THAT'S LIFE:
WHY YOU AND YOUR iPOD MUST DIE

Seth Cook is the oldest living American with a particularly rare genetic disorder. He's lost all his hair. His skin is covered in wrinkles. His arteries are hardened. His joints hurt from arthritis. He takes an aspirin and a blood thinner every day.

He is twelve years old.

Seth has Hutchinson-Gilford progeria syndrome, often just called progeria. Progeria is very rare—thought to occur in just 1 of every 4 to 8 million births. It's also very unfair; the word comes from the Greek for prematurely old, and that's the difficult fate in store for people born with it. Children who have progeria age at up to ten times the speed of people without it. By the time a baby who has progeria is about a year and a half old, his or her skin starts to wrinkle and their hair starts to fall out. Cardiovascular problems, like hardening of the arteries, and degenerative diseases, like arthritis, soon follow. Most people who have progeria die in their teens of a heart attack or a stroke; nobody is known to have lived past thirty.

Hutchinson-Gilford progeria isn't the only disease that causes accelerated aging—it's just the most heartbreaking, because it's the fastest, and it starts at birth. Another aging disorder, Werner syndrome, doesn't manifest itself until someone carrying the mutation that causes it reaches puberty; it's sometimes called adult-onset progeria. After puberty, rapid aging sets in, and people who have Werner syndrome usually die of age-related disease by their early fifties. Werner syndrome, although more common than Hutchinson-Gilford progeria, is still very rare, affecting just one in a million.

Because these rapid-aging diseases are so uncommon, they haven't been the focus of much research (and they're called orphan diseases for that reason). But that's starting to change, as scientists have realized that they hold clues about the normal aging process. In April 2003, researchers announced that they had isolated the genetic mutation that causes progeria. The mutation occurs in a gene that is responsible for the production of a protein called lamin A. Normally, lamin A provides structural support for the nuclear membrane, the package that houses your genes at the core of every cell. Lamin A is like the rods that hold up a tent—the nuclear membrane is organized around it and supported by it. In people who have progeria, lamin A is defective and cells deteriorate much more rapidly.

In 2006, a different team of researchers established a link between lamin A deterioration and normal human aging. Tom

Misteli and Paola Scaffidi, researchers at the National Institutes of Health, reported in *Science* that the cells of normal elderly people show the same kinds of defects that are found in the cells of people who have progeria. That's very significant—it's the first confirmation that the accelerated aging that characterizes progeria is related to normal human aging on a genetic level.

The implications are far-reaching. More or less since Darwin described adaptation, natural selection, and evolution, scientists have been debating where aging fits into the picture. Is it just wear and tear, the way your favorite shirt picks up little stains and rips and marks over the years, eventually fraying and wearing out? Or is it the product of evolution? In other words, is aging accidental or intentional?

Progeria and the other accelerated-aging diseases suggest that aging is preprogrammed, that it's part of the design. Think about it—if a single genetic error can trigger accelerated aging in a baby or an adolescent, then aging can't only be caused by a lifetime of wear and tear. The very *existence* of the progeria gene demonstrates that there could be genetic controls for aging. That, of course, raises a question you've no doubt come to expect. Are we programmed to die?

LEONARD HAYFLICK IS one of the fathers of modern aging research. During the 1960s he discovered that (with one special exception) cells only divide a fixed number of times before they up and quit. This limit on cellular reproduction is appropriately called the Hayflick limit; in humans the limit is around fifty-two to sixty.

The Hayflick limit is related to the loss of a genetic buffer at the end of chromosomes called telomeres. Every time a cell reproduces it loses a little bit of DNA. In order to prevent that information loss from making a difference, your chromosomes have what amounts to extra information at their tips; those bits of information are telomeres.

Imagine you have a manuscript and need to make fifty copies but Kinko's has just thrown you a curveball. Instead of charging you money, they're just going to take one page off the end of your manuscript after every copy. That's a problem—your manuscript is two hundred pages long; if you give them a page after every copy, the last copy is only going to have one hundred fifty pages and whoever gets it is going to miss a quarter of the story. So, being a highly evolved organism with a gift for clever solutions, you add fifty blank pages to the end of your manuscript and present Kinko's with a two-hundred-fifty-page manuscript. Now, all fifty copies will have the complete story; you won't lose a page of precious information until you decide to make copy fifty-one. Telomeres are like blank pages; as cells reproduce, telomeres are shortened, and the truly valuable DNA is protected. But once a cell replicates between fifty and sixty times, the telomeres are essentially gone and the good stuff is in jeopardy.

Now, why would we evolve a limit against cellular reproduction?

In a word? Cancer.

IF THERE'S A health-related word more closely associated with fear and mortality than cancer, I don't know what it is. It's so widely assumed to be a likely death sentence that, in millions of families, it's barely spoken out loud; instead it's only spoken, if at all, in a kind of stage whisper.

As you no doubt know, cancer isn't a specific disease; it's a family of diseases characterized by cell growth gone haywire. And the truth is, some cancers are highly treatable—many of them have higher survival rates and better chances for complete recovery than other common health problems, such as heart attacks and strokes.

As we've discussed, your body has multiple lines of cancer defense. There are specific genes responsible for tumor suppression. There are genes responsible for creating specialized cancer hunters programmed to seek and destroy cancer cells. There are genes responsible for repairing the genes that fight cancer. Cells even have a mechanism to commit a kind of hara-kiri. Apoptosis, or programmed cell death, occurs when a cell detects that it has become infected or damaged—or when other cells detect a problem, and "convince" the dangerous cell to kill itself. And on top of that there's the Hayflick limit.

The Hayflick limit is a potent check against cancer—if everything goes wrong in a cell and it becomes cancerous, the Hayflick limit still prevents its unchecked reproduction, essentially shutting down tumor growth before it really gets going. If a cell can only reproduce a specific number of times before it runs out of steam, it can't reproduce uncontrollably, right?

Right—as far as it goes. The problem is, cancer cells are sneaky little villains with a few tricks up their cellular sleeves. One of those is an enzyme called telomerase. Remember that the Hayflick limit works through telomeres—when they run out, cells die or lose the ability to reproduce. So what does telomerase do? It lengthens those telomeres at the ends of chromosomes. In normal cells telomerase is usually not active and therefore telomeres are usually shortened. But cancer cells can sometimes kick telomerase into high gear, so that the telomeres are replenished more rapidly. When that happens, there's less loss of genetic information, because the telomere buffer never runs out. The expiration date programmed into cells is canceled, and the cell can reproduce forever.

When cancer cells are successful, it's usually with the help of telomerase. More than 90 percent of the cells in cancerous human tumors use telomerase. That's how they become tumors—without telomerase, cancer cells would die out after dividing fifty to sixty times, or perhaps a little longer. With telomerase helping them to short-circuit the Hayflick limit, they can multiply uncontrollably, wreaking the biological havoc we're all too familiar with. On top of all that, successful cancer cells—the cells we most want to die on their own—have found a way around apoptosis, or programmed cell death. They ignore the suicide command that noncancerous cells obey when they become infected or damaged. In biological terms, that makes cancer cells "immortal"—they can divide forever. Scientists are currently working to perfect a test that detects increased telomerase activity; that could give doctors a powerful new tool to help reveal hidden cancer cells.

The other exceptions to the Hayflick limit, by the way, are those current stars of political, medical, and ethical debate—stem cells. Stem cells are "undifferentiated" cells—in other words, they can divide into many different kinds of cells. A B-cell that makes your antibodies can only divide into another B-cell, and a skin cell can only produce another skin cell. Stem cells can produce many types of cells—the mother of all stem cells, of course, is the single cell that started you off in your mother. A zygote (which is the union of a sperm and an egg) obviously has to be able to produce every kind of cell; otherwise you'd still be a zygote. Stem cells are not subject to the Hayflick limit—they're also immortal. They pull off this feat by using telomerase to fix their telomeres the same way that some cancer cells do. You can see why scientists believe stem cells have such potential to cure disease and alleviate suffering—they have the potential to become anything and they never run out of steam.

Many scientists believe cancer prevention is the "reason" cells have evolved with a limit on the number of times they can reproduce. The flip side to the Hayflick limit, of course—compromise, compromise—is aging. Once cells hit the limit, future reproductions don't really work and things start to break down.

CANCER PROTECTION AND the Hayflick limit aren't the only evolutionary explanations for the aging mechanism. First of all, that doesn't necessarily explain why different animals—even closely related ones—have such different life expectancies.

It's interesting to note that, in mammals, with a few exceptions, there's a close correlation between size and life expectancy. The bigger you are, the longer you live. (That doesn't mean you should head to Dairy Queen—the bigger the natural size of the *species*, the longer the average member of the species lives, not the bigger the individual.) The longer life expectancy of larger mammals is at least partially due to their superior ability to repair DNA. But that explains, at least in part, *how* we live longer; it doesn't explain *why* we big creatures developed those superior repair mechanisms.

One theory suggests that there is a direct connection between shorter life expectancy and greater external threats. I'm not just saying that the risk of being eaten reduces an animal's life expectancy, although it does, of course. Essentially, animals with a greater risk of being eaten evolve to live shorter lives—even if they aren't eaten. Here's how—if a species faces significant environmental threats and predators, it's under greater evolutionary pressure to reproduce at an early age, so it evolves to reach adulthood faster. (A shorter life span also means a shorter length of time between generations, which allows a species to evolve faster—which is important for species that face a lot of environmental threats; that's one of the things that helps rodents develop resistance to poisons relatively quickly.) At the same time, there's never any real evolutionary pressure to evolve mechanisms to repair DNA errors that occur over time, because most individuals in the species don't live long enough to experience those errors. You wouldn't buy an extended warranty on an iPod if you were only going to keep it for a week. On the flip side, a species that is more dominant in its environment, and that can continue to reproduce for most of its life, will gain an advantage in repairing accumulated DNA errors. If it lives longer, it can reproduce more.

I believe that programmed aging confers an evolutionary benefit on the *species,* not the individual. According to this thinking, aging acts like a biological version of planned obsolescence. Planned obsolescence is the often denied but never disproved notion that manufacturers of everything from refrigerators to cars build a shelf life into their products, essentially guaranteeing that they wear out after a limited number of years. This does two things—one arguably to the consumer's benefit, the other certainly to the manufacturer's benefit. First, it makes the way for new, improved versions. Second, it means you need to buy a new fridge. Some people accused Apple of employing planned obsolescence in the development of its superpopular iPods a few years ago—manufacturing them with batteries that only lasted for about eighteen months and couldn't be replaced, forcing consumers to buy a new model when their battery died. (Apple now has a battery replacement program, although it's tantamount to an iPod replacement program—for a small fee, they send you a new or refurbished equivalent to your now-powerless purchase.)

Biogenic obsolescence—that is to say, aging—might accomplish two similar ends. First, by clearing out older models, aging makes room for new models, which is exactly what creates the room for change—for evolution. Second, aging can protect the group by eliminating individuals that have become laden with parasites, preventing them from infecting the next generation. Sex and reproduction, in turn, are the way a species gets upgraded.

THE PROSPECT OF programmed aging opens up the door to all kinds of exciting possibilities. Already, scientists are exploring benefits that may be found by turning aging mechanisms off— and by turning them back on. The possibility of short-circuiting telomerase in cancer cells—the enzyme that cancer cells use to make themselves immortal—may lead to powerful new weapons against cancer.

A year before they did so, the researchers who first linked progeria-related aging to normal aging also demonstrated that it is possible to reverse the cellular damage caused by progeria. They applied a "molecular Band-Aid" to progeria cells in their lab and eliminated the defective lamin A. After a week, more than 90 percent of the cells they treated looked normal. They haven't been able to reverse progeria in people yet, but every new insight is a step in the right direction. The combined implication of the two studies isn't exactly a map to Ponce de León's fabled fountain of youth, but it's certainly intriguing. Cells in aging humans are programmed to break down in a similar fashion to progeria cells. And scientists have been able to reverse those breakdowns in the lab. The operative words in the last two sentences?

Aging. And *reverse.* Now *that's* something to look forward to.

Speaking of things to look forward to, this book is all about life. About why we are who we are and why we work the way we do. And there's one place where all of that really comes together—evolution's ultimate laboratory—the womb.

CONGRATULATIONS! YOU'RE HAVING a baby!

Over the next nine months, millions of years of interaction with disease, parasites, plagues, ice ages, heat waves, and countless other evolutionary pressures—not to mention a little romance—will come together in a stunningly complex interaction of genetic information, cellular reproduction, methyl marking, and the commingling of germ lines to produce your little peanut.

You and your partner are doing the evolution dance, contributing eons of genetic history to the next generation. It's an amazing, uplifting, deeply moving process. Which is why you should be forgiven when you go to the hospital to have your baby and feel a little put off by the surroundings—just about everybody in the place is sick, trying to ward off disease or death, and you're there to bring a little life into the world.

You look at the directory to find out where to go and you read something like

CARDIOLOGY
ENDOCRINOLOGY

GASTROENTEROLOGY

GENERAL SURGERY

You skip ahead and read

HEMATOLOGY

INFECTIOUS DISEASES

INTENSIVE CARE UNIT (ICU)

LABORATORY MEDICINE AND PATHOLOGY

And then, finally, there it is—Obstetrics and Gynecology—sandwiched right between those two heartwarmers Neurosurgery and Psychiatry.

Soon you will be hustled upstairs, hurried into a hospital gown, and hooked up to an IV; if you've ever been to a hospital before because you were actually *sick*—instead of pregnant—it's all probably feeling a bit too familiar right about now. You're having a baby—couldn't they make it a little more fun?

Of course, all of the medical drama is for very good reason; in 2000 the United Nations estimated that more than half a million mothers died of complications resulting from pregnancy—but less than 1 percent of those deaths were in the developed world. So there's no question that modern medicine has helped to remove the great portion of risk from childbirth. But the approach tends to be one that is sort of disease-oriented—usually treating pregnancy as a risk to be managed, rather than an evolutionary miracle that just needs to be helped along.

Perhaps our ability to make pregnancy and childbirth even more safe and comfortable would benefit by asking the same questions we're starting to ask about our relationship to disease. Why has evolution led humans to give birth the way we do?

CHILDBIRTH IN HUMANS is riskier, is longer, and certainly seems more painful than it is in any of our genetic cousins. Ultimately, that can be traced to two things—crossword puzzles and marching bands. Well, maybe not crossword puzzles and marching bands per se, but it *is* because of the two characteristically human traits that allow us to do them—big brains and bipedalism. When it comes to birth, those two traits are a tricky combination.

The skeletal adaptations that allow us to walk on two feet changed the structure of the human pelvis—unlike the pelvis of monkeys, apes, and chimps, the human pelvis regularly has to bear the weight of your entire upper body. (Chimps *do* walk on two legs from time to time, but usually only to carry food or wade across rivers and streams.) The evolution toward bipedalism included selection of a specialized pelvis that makes walking upright possible—which in true evolutionary style came with a compromise. According to Wenda Trevathan, a biological anthropologist who has spent much of her career studying the evolution of birth, the human pelvis is "twisted" in the middle; it starts off pretty wide, and is broad from side to side at the birth canal's "entrance," but gets narrower as it goes on, ending in an "exit" that presents a pretty tight squeeze for an infant's skull.

Millions of years after we learned to walk on two feet, we started evolving bigger brains. Bigger brains need bigger skulls. And eventually (after a few million years, that is) human women with small birth canals were giving birth to human babies with big skulls. That, by the way, is one of the reasons why a newborn's head is so vulnerable—the skull is actually composed of separate plates connected by tissue called sutures that give it the flexibility to squeeze through the birth canal. The plates don't start fusing together until the baby is about twelve to eighteen months old, and they don't become fully fused until adulthood (much later than in chimps).

The big brain is so difficult to get out of the tight birth canal that most of human brain development takes place after birth. When monkeys are born, their brains are more than 65 percent of the size that they'll be when fully grown. But baby human brains are only 25 percent of the size—that's one reason babies are so helpless for the first three months; their brains are in a state of rapid development. Many doctors actually call it the fourth trimester.

On top of all that, the human birth canal isn't one constant shape, so the fetus has to twist its way through. When it does emerge, it's usually facing away from its mother because of all that twisting, adding one more difficulty to human birth. Chimps and monkeys come out facing their mothers. Imagine a mother chimp squatting during delivery and the baby chimp emerging from the birth canal facing upward toward its mother and you've got a pretty good picture. The mother chimp can reach down, cradle the infant's head from behind its neck, and help with its delivery. In humans, the mother can't do that (even if she is squatting) because the baby is facing away—if she tries to assist the baby she risks bending its neck or spine the wrong way and causing serious injury. Trevathan believes this "triple threat" of big brains, a pelvis designed for walking, and backward-facing babies led to the nearly universal human tradition of helping one another with delivery. Every other primate generally goes it alone when it comes time to give birth.

If you pause and think about this for a moment in light of everything we know about evolutionary pressure, it's a little confusing. Why would evolution favor adaptations that made reproduction more dangerous? Well, it wouldn't—unless it made survival so much more likely that it outweighed the increased reproductive risk. For example, if an adaptation allowed twice as many babies to reach adulthood and get pregnant, it might be worth the risk that a small percentage of them wouldn't survive childbirth.

It's pretty clear that big brains are a big advantage. But what about walking upright? Why did we evolve in that direction? Why aren't we a bunch of smart hominids crawling to the grocery store on all fours or swinging to the library through the trees instead of strolling along a sidewalk?

Something clearly sent our human ancestors off in a different evolutionary direction from the one followed by the ancestors of the modern chimp or ape. Whatever it was, it ultimately prompted a cascade of evolutionary dominoes, with one adaptation leading to another. As a writer named Elaine Morgan (whom we'll hear more from shortly) put it, "Our ancestors entered the Pliocene [a geological time scale about 2 to 5 million years ago] as hairy quadrupeds with no language and left it hairless, upright and discussing what kinds of bananas they liked best." And that's not all. We also became fatter, developed prominent noses with nostrils pointing downward, and lost much of our sense of smell.

So what happened?

THE CONVENTIONAL WISDOM about our shift from all fours to two feet is the "savanna hypothesis." The savanna theory holds that our apelike ancestors abandoned the dark African forests and moved into the great grassy plains, perhaps because of climate changes that led to massive environmental change. In the forest, food was plentiful—fruits, nuts, and leaves could be found in abundance. But out in the savanna, life was tougher, so the theory goes, and our ancestors had to find new ways to get food. Males began to hunt bravely for meat among the herds of grazing animals. Some combination of these new circumstances—the need to scan the horizon for food or predators, the need to cover long distances between food and water—led the savanna hominid to begin walking upright. Other adaptations were similarly related to the new environment—hunting required tools and cooperation; smarter prehumans made better tools and better teammates, so they survived longer and attracted more mates, and the process selected for bigger brains. The savanna was hot, and all those brave males chasing animals tended to overheat, so they lost their hair to keep them cool.

That's the conventional theory, anyway.

But Elaine Morgan isn't a conventionalist, and she isn't buying it. Morgan is a prolific Welsh writer who originally became interested in evolution more than thirty years ago. As she read books describing the savanna theory, she was immediately skeptical. For starters, she couldn't understand why evolution—so concerned with reproduction—would be driven only by the requirements of the male. "The whole thing was very focused on the male," she recalls. "Their premise was that the important thing was the evolution of man-the-hunter. I began to think: 'They must have this wrong.'" Shouldn't evolution be at least as influenced toward women and children?

In a word?

Yes.

BY THE TIME Morgan was questioning it, the savanna hypothesis was well entrenched in the scientific community. And like most well-entrenched theories, those who challenged it were generally ignored or ridiculed. But that wasn't the kind of thing to stop Elaine Morgan. So, certain that the savanna theory's men-only approach to evolution didn't make sense, Morgan set out to write a book exposing its flaws. It wasn't intended to be a scientific book; rather, she attacked the savanna theory with that ancient and highly effective debunker of all things highfalutin—common sense.

The Descent of Woman was published in 1972, and it roundly savaged the idea that male behavior was the driving force in human evolution. Humans started walking on two legs so we could cover distances between water and food *faster* than we could on four legs? Yeah, right—ever race a cheetah? Even some of the slower quadrupeds can outrun us. We lost our hair because the *males* got too hot chasing antelope? So why do females have even less hair than males? And what about all those other hairless animals running around the savanna? Oh, right, there aren't any. Every hairless mammal is aquatic or at least plays in the mud—think of hippos, elephants, and the African warthog. But there aren't any hairless primates. In researching her book, Morgan came across the work of a marine biologist named Alister Hardy. In 1960, Hardy offered a different theory to explain our evolutionary divergence from other primates. He suggested that a band of woodland apes became isolated on a large island around what is now Ethiopia and adapted to the water, regularly wading, swimming, and foraging for food in lagoons. Hardy first got the idea nearly thirty years earlier when reading a book by Professor Wood Jones, called *Man's Place among the Mammals,* which asked why humans were the only land mammals with fat attached to our skin. Pinch your dog or cat and you'll feel the difference when you grab a fistful of nothing but skin. Hardy was a marine biologist; he made an immediate connection to marine mammals—like hippos, sea lions, and whales—all of which have fat directly attached to the skin. He figured there could only be one reason for humans to share a trait that was otherwise only found in aquatic or semiaquatic mammals—an aquatic or semiaquatic past.

An aquatic ape.

Nobody took Hardy's theory seriously, not even seriously enough to challenge it. Until Elaine Morgan came along. And she took it seriously enough to write five books about it—so far.

Morgan builds a compelling case. Here's the essence of the aquatic ape hypothesis, as it's now known. For a long stretch of time, our prehuman ancestors spent time in and around the water. They caught fish and learned to hold their breath for long periods while diving for food. Their ability to survive on land and water gave them twice as many options to avoid predators as their land-bound cousins—chased by a leopard, the semiaquatic ape could dive into the water; chased by a crocodile, it could run into the forest. Apes that spent time in the water would naturally evolve toward bipedalism—standing upright allowed them to venture into deeper water and still breathe, and the water helped to support their upper bodies, making it easier for their bodies to support them on two feet.

The aquatic ape theory explained why, like many other aquatic mammals, we lost our fur—to become more streamlined in the water. It explained the development of our prominent nose and downward-facing nostrils, which allowed us to dive. The only other primate with a prominent nose (that we know of) is the aptly named proboscis monkey—which just so happens to be semi-aquatic itself and can also be seen wading in the water on two legs or going for a swim.

Finally, the aquatic theory may explain why our fat is attached to our skin. Like other aquatic mammals, such as dolphins and seals, it allows us to flow smoothly through the water using less energy. Human babies are also born with significantly more fat than baby chimps or monkeys. Providing all that fat is an additional burden to the mother, so there's got to be a good reason for it. Most scientists agree it helps to keep the baby warm. (Remember brown fat? The special heat-generating fat that is usually only found in human newborns?) Elaine Morgan thinks that besides keeping babies warm, the extra fat also helps to keep them afloat. Fat is less dense than muscle, so a higher percentage of body fat makes people more buoyant.

The debate over the semiaquatic ape is far from over. Most mainstream anthropologists certainly still subscribe to the savanna hypothesis. And the semiaquatic versus savanna smackdown tends to provoke emotion on both sides that makes it harder to resolve. One of the things that get lost in the scientific shouting is just what the aquatic ape hypothesis actually holds. It doesn't suggest that there was some prehuman animal that lived mostly under-water and only surfaced periodically for air like some kind of primate whale. A British computer programmer named Algis Kuliukas read Morgan's work after his wife gave birth in a birthing tub. He was shocked to find that many of the scholars who railed against Morgan's theory freely acknowledged the possibility that human ancestors spent time in the water and that their time in the water could have influenced evolution. If they acknowledged that, what was all the fuss about?

Kuliukas realized a good deal of the controversy over the theory was related to a lack of understanding over just what the theory actually held. He wrote:

[Some critics] . . . never really "got" what the theory was. They think they have—but they're just wrong. They think it's suggesting that humans went through some "phase" of almost becoming mermaids or something and they reject it as nonsense on that basis.

So Kuliukas decided to try and add a little clarity to the conversation by proposing a simple summation of the aquatic ape hypothesis:

That water has acted as an agent of selection in the evolution of humans more than it has in the evolution of our ape cousins. And that, as a result, many of the major physical differences between humans and the other apes are best explained as adaptations to moving (e.g. wading, swimming and/ or diving) better through various aquatic media and from greater feeding on resources that might be procured from such habitats.

When you put it like that, it starts to sound an awful lot like common sense, don't you think?

LET'S IMAGINE THAT Alister, Elaine, and Algis are right. Some of our ancestors spent a lot of time in and around the water, so much so that it influenced our evolution. And let's further assume that it was in this environment that we first learned to stand on our own two feet. That, in turn, allowed for the change to our pelvis and twisted the birth canal, making childbirth more difficult. So that means the first bipedal childbirths might have been of semi-aquatic apes in a semiaquatic environment.

That still doesn't explain the lack of evolutionary pressure *against* bipedalism and the accompanying reproductive risk caused by the change in pelvic shape. Unless—what if the water changed the equation somehow and made the process easier? If the water made the birthing process easier, then most of the evolutionary pressure would favor the advantages those aquatic apes gained from the shift to two feet.

But if the water made it easier for aquatic apes with small pelvic openings to give birth, then shouldn't water make it easier for humans with small pelvic openings to give birth?

LEGEND HAS IT that the first medical water birth took place in the early nineteenth century in France. Birth attendants were struggling to help a woman who had been in labor for more than forty-eight hours when one of the midwives suggested a warm bath might help the expectant mother to relax. According to the story, the baby was born shortly after the woman settled into the tub.

A Russian researcher named Igor Tjarkovsky is often credited as the father of modern water birthing. He designed a special tank in the 1960s for water birthing, but the trend didn't really catch on in the West until the early 1980s or so. The reaction of the medical establishment wasn't encouraging. In medical journals and the popular press, doctors suggested that water birthing was dangerous, filled with unacceptable risks of infection and drowning. It wasn't until 1999, when Ruth Gilbert and Pat Tookey of the Institute of Child Health in London published a serious study showing that water birth was at least as safe as conventional methods, that all these predictions of doom and gloom were shown to be largely baseless.

An even more recent Italian study, published in 2005, has confirmed the safety of water birthing—and demonstrated some stunning advantages. The Italian researchers compared 1,600 water births at a single institution over eight years to the conventional births at the same place during the same time.

First of all, there was no increase of infection in either mothers or newborns. In fact, there was apparently an additional protection for the newborn against aspiration pneumonia. Babies don't gasp for air until they feel air on their face; when they're underwater, the mammalian diving reflex—present in all mammals—triggers them to hold their breath. (Fetuses do "breathe" while in their mother's womb, but they're actually sucking in amniotic fluid, not air, which forms a crucial part of their lung development.) When babies are delivered conventionally, they take their first breath of air as soon as they feel air on their face; sometimes, if they get in a big breath before the doctor can clean their face, this causes them to inhale fecal matter or "birthing residue" that can cause an infection in their lungs—aspiration pneumonia. But babies delivered underwater don't face that risk—until they're brought to the surface they don't switch from fetal circulation to regular circulation, so there's no risk of them inhaling water, and attendants have plenty of time to clean their faces while they're still underwater, before lifting them out of it and triggering their first breath.

The study revealed many more benefits. First-time mothers delivering in water had a much shorter first stage of labor. Whether the water relaxed nervous minds or tired muscles or had some other effect, it clearly accelerated the delivery process. Women delivering in water also had a dramatic reduction in the need for episiotomies—the surgical cut routinely performed in hospital births to expand a woman's vaginal opening in order to prevent complications from tearing. Most of the time they just weren't necessary—the water simply allowed for more of a stretch.

And perhaps most remarkably, the vast majority of the women who gave birth in water needed no painkillers. Only 5 percent of the women who started their labor in water asked for an epidural—compared to 66 percent of the women who gave birth through conventional means.

The behavior of human newborns in the water offers another tantalizing suggestion that the aquatic ape theory holds water. A child development researcher named Myrtle McGraw documented these surprising abilities back in 1939—not only do very young babies reflexively hold their breath, they also make rhythmic movements that propel them through the water. Dr. McGraw found that this "water-friendly" behavior is instinctual and lasts until babies are about four months old, when the movements become less organized.

Primitive swimming would be an awfully surprising instinct for an animal that evolved into its more or less current form on the hot, dry plains of the African savanna. Especially when that animal is born relatively helpless, with almost no other instinctual behavior besides eating, sleeping, and breathing.

And crying. Can't forget crying. Of course, if you *are* having a baby, you won't.

GIVE YOUR BABY a few years and he or she will trade in the cries for whys. Why do I have to go to bed? Why do you have to go to work? Why can't I have dessert for breakfast? Why does my stomach hurt? Why?

You tell your toddler to keep the questions coming. That's what this book is all about. Questions. Two in particular, many times over. The first is, "Why?"

Why do so many Europeans inherit a genetic disorder that fills their organs with iron?

Why do the great majority of people with Type 1 diabetes come from Northern Europe?

Why does malaria want us in bed but the common cold want us at work?

Why do we have so much DNA that doesn't seem to do anything?

The second question, of course, is, "What can we do with that?"

What can we do with the idea that hemochromatosis protected people from the plague?

What can we do with the possibility that diabetes was an adaptation to the last ice age?

What does it mean for me to understand that malaria wants me laid up and the cold wants me on the move to help them each spread?

And what does it mean that we have all this genetic code that probably came from viruses and sometimes jumps around our genome?

Oh, not much.

Just develop new ways to combat infection by limiting bacterial access to iron and provide better treatment to people whose iron deficiencies are actually natural defenses against highly infectious environments.

Just open up exciting new avenues of research by leading us to explore animals, like the wood frog, that use high blood sugar to survive the cold and manage it successfully.

Just lead us to search for ways to direct the evolution of infectious agents *away* from virulence and toward harmlessness— instead of waging an antibiotic war that we may never be able to win.

Just . . . who knows?

If we don't ask, we'll never find out.

CONCLUSION

I hope that you'll come away from this book with an appreciation of three things. First, that life is in a constant state of creation. Evolution isn't over—it's all around you, changing as we go. Second, that nothing in our world exists in isolation. We—meaning humans and animals and plants and microbes and everything else— are all evolving together. And third, that our relationship with disease is often much more complex than we may have previously realized.

Life after all is a complicated gift—an almost impossible assemblage of biology, chemistry, electricity, and engineering that adds up to a miraculous whole so much greater than the sum of its parts. The entire universe is geared toward disorder. Given all the forces pulling for disorder, it's a wonder that we live at all—and as long and as well as most of us do. Which is why, instead of taking our health for granted, we should appreciate it with the reverence it deserves.

When you make that mental leap—when you think of the amazing gift of your health and your life in the context of all the nearly incomprehensible forces of the universe pulling toward chaos—it reorients you, imbuing you with a deep respect for the immensely beautiful and intricate design of life on earth. Life that has been created and re-created again and again through billions of years of trial and toil. Something so complicated and time-consuming that it has to be a labor of love.

The more we learn about the unbelievably complex, immensely varied, and yet simultaneously simple origin and development of life on earth, the more it looks like a miracle, and one that is still unfolding.

The miracle of evolution.

ACKNOWLEDGMENTS

I am indebted to Massey College, at the University of Toronto, for providing support for a sabbatical and the rich interdisciplinary environment where many of the ideas found in this book were formed and nurtured. My gratitude to Master John Fraser and John Neary for making my fellowship at Massey possible and so enjoyable. I must thank all the special people at Mount Sinai School of Medicine in New York, who went above and beyond the call of duty, making all the arrangements to allow me the time necessary to finish writing. Many thanks to my very dedicated research assistants, the incredible fact-checking duo of Richard Verver and Ashley Zauderer. I am grateful to all the scientists I have collaborated with over the years, and to all the researchers and support personnel who have come before me, without whom there would be no research to write about. I have been blessed with many special mentors, in particular my dear friends Maire E. Percy, Gard W. Otis, Katherine Elliott, and Daniel P. Perl. I'm indebted to Claire Wachtel at HarperCollins, not only for her friendship and continual honesty but also for ferociously adopting this book and never letting a wet nurse near it. Michael Morrison, David Roth-Ey, Lynn Grady, and Lisa Gallagher believed in this project from the beginning, and Dee Dee DeBartlo has done a superb job of making sure the world knows about this book. Thanks to Kim Lewis for her patience shepherding this through production, and Lauretta Charlton, who has been a lifesaver. I am grateful to my agent, Dorian Karchmar of the William Morris Agency, for tirelessly championing this project from start to finish and coming up with the title. Thanks to Tracy Fisher and Raffaella De Angelis, both of William Morris, and Shana Kelly at William Morris (London) for taking the book global, and to Andy McNicol, who took care of audio rights. To the entire team at William Morrow, your efforts did not go unnoticed. And finally, I have to thank Jonathan Prince, whose inspired writing helped to elevate this project immensely.

NOTES

Much of the scientific foundation of this book was drawn from my research and that of my collaborators. I have also made liberal use of the work of other scientists, including unpublished research and personal interviews. I hope that the notes below will give you a detailed understanding of the sources, and also provide a launching point for you to find more information about the topics discussed.

INTRODUCTION

ix   my grandfather and me

Ann McIlroy, "Teenager Sharon Moalem Suspected His Grandfather's Alzheimer's Was Linked to a Buildup of Iron in His Brain. Years Later, He Proved It," *Globe and Mail,* January 31, 2004.

xii   checking for hemochromatosis

The blood tests mentioned in the introduction as a screen for hemochromatosis include the following: total iron binding capacity (TIBC), serum iron, ferritin, and % transferrin saturation. There is also a commercial genetic test available (these can be quite expensive) for the presence of hemochromatosis mutations, but I would not recommend having the test done until there is robust legislation that protects individuals from genetic discrimination.

xii   evolution and medicine

E. R. Stiehm. 2006. Disease versus disease: how one disease may ameliorate another. *Pediatrics* 117(1):184–191; Randolph M. Nesse and George C. Williams, "Evolution and the Origins of Disease," *Scientific American*, November 1998; R. M. Nesse. 2001. On the difficulty of defining disease: a Darwinian perspective. *Med Health Care Philos* 4(1):37–46; E. E. Harris and A. A. Malyango. 2005. Evolutionary explanations in medical and health profession courses: are you answering your students' "why" questions? *BMC Med Educ* 5(1):16.

xiv   *you are not alone*

S. R. Gill, M. Pop, R. T. Deboy, et al. 2006. Metagenomic analysis of the human distal gut microbiome. *Science* 312(5778):1355–1359.

xv   *DNA isn't destiny*

See pages 183–198 in Lenny Moss, *What Genes Can't Do* (Cambridge, MA: MIT Press, 2003); pages 8–47 in Michael Morange, *The Misunderstood Gene* (Cambridge, MA: Harvard University Press, 2001); H. Pearson. 2006. Genetics: what is a gene? *Nature* 441(7092):398–401.

CHAPTER I: IRONING IT OUT

1   Aran Gordon and hemochromatosis

Kathleen Johnston Jarboe, "Baltimore Business Executive Runs for His Life and Lives of Others," *The Daily Record*, April 22, 2005. For a good resource book on hemochromatosis see C. D. Garrison, Iron Disorders Institute, *The Iron Disorders Institute Guide to Hemochromatosis* (Nashville, TN: Cumberland House, 2001). To watch an NBC news interview with Aran visit www .iron disorders.org/Aran/.

5   *the Geritol Solution*

F. M. Morel and N. M. Price. 2003. The biogeochemical cycles of trace metals in the oceans. *Science* 300(5621):944–947; D. J. Erickson III and J. L. Hernandez. 2003. Atmospheric iron delivery and surface ocean biological activity in the Southern Ocean and Patagonian region. *GeoPhys Res Lett* 30(12):1609–1612; J. H. Martin, K. H. Coale, K. S. Johnson, et al. 2002. Testing the iron hypothesis in ecosystems of the equatorial Pacific Ocean. *Nature* 371:123–129; Richard Monastersky, "Iron versus the Greenhouse," *Science News,* September 30, 1995; Charles Graeber, "Dumping Iron," *Wired,* November 2000.

6   Eugene D. Weinberg and his lifelong infatuation with iron

To get it directly from the source see E. D. Weinberg and C. D. Garrison, *Exposing the Hidden Dangers of Iron: What Every Medical Professional Should Know about the Impact of Iron on the Disease Process* (Nashville, TN: Cumberland House, 2004).

9   bubonic plague

N. E. Cantor, *In the Wake of the Plague: The Black Death and the World It Made* (New York: Perennial/HarperCollins, 2002); J. Kelly, *The Great Mortality: An Intimate History of the Black Death, the Most Devastating Plague of All Time* (New York: HarperCollins, 2005).

9   *Alas! Our ships enter the port*

Gabriele de'Mussi, *Istoria de morbo siue mortalitate que fuit de 1348,* page 76 in G. Deaux, *The Black Death, 1347* (New York: Weybright and Talley, 1969). For more about the plague as literary motif see www.brown.edu/Departments/ Italian_Studies/dweb/plague/perspectives/de_mussi.shtml.

10   Passover and the plague

For an intriguing account of the possible connections between the Jewish observances of Passover and plague prevention see M. J. Blaser. 1998. Passover and plague. *Perspect Biol Med* 41(2):243–256.

11   *Father abandoned child*

Angelo di Tura, Seina Chronicle, 1354, pages 13–14 in W. M. Bowsky, *The Black Death: A Turning Point in History* (New York: Holt, 1971).

11   *"Iron status mirror[ed] mortality"*

In some of the most recent outbreaks of plague men and women seemed to be equally affected. This could possibly be attributed to our diets' being more iron "rich" as a result of fortification of grains and processed foods. See S. R. Ell. 1985. Iron in two seventeenth-century plague epidemics. *J Interdiscip Hist* 15(3):445–457, which provides more on the epidemiology of bubonic plague and how young men seem to have been most susceptible.

12   plague in London

For a great article by Graham Twigg with maps of parishes during the plague in London see www.history.ac.uk/cmh/epitwig.html, originally published as *Plague in London: Spatial and Temporal Aspects of Mortality,* in *Epidemic Disease in London,* ed. J. A. I. Champion, Centre for Metropolitan History Working Papers Series, No. 1 (1993).

12   hemochromatosis and the plague

For the original paper describing the proposed connection between hemochromatosis and the plague see S. Moalem, M. E. Percy, T. P. Kruck, and R. R. Gelbart. 2002. Epidemic pathogenic selection: an explanation for hereditary hemochromatosis? *Med Hypotheses* 59(3):325–329. For more information on the importance of iron in bacterial infections see S. Moalem, E. D. Weinberg, and M. E. Percy. 2004. Hemochromatosis and the enigma of misplaced iron: implications for infectious disease and survival. *Biometals* 17(2):135–139.

13   *the Bruce Lees of the immune system*

Researchers have yet to test the fighting ability of macrophages of people who have hemochromatosis directly. Yet in a recent study researchers found that the microbe that causes tuberculosis *(Mycobacterium tuberculosis)* had a much more difficult time acquiring iron from cells of people who had hemochromatosis. Since most pathogenic bacteria (like *Yersinia pestis,* which is thought to have caused the bubonic plagues) and fungi depend upon iron for their infectivity, it is thought that this may in fact be the advantage that led hemochromatosis mutations to become so prevalent in Western Europe. The following reference is for the experiments mentioned: O. Olakanmi, L.S. Schlesinger, and B. E. Britigan. 2006. Hereditary hemochromatosis results in decreased iron acquisition and growth by *Mycobacterium tuberculosis* with human macrophages. *J Leokoc Biol* (Epub October 12, 2006, ahead of print); O. Olakanmi, L. S. Schlesinger, A. Ahmed, and B. E. Britigan. 2002. Intraphagosomal *My-*

cobacterium tuberculosis acquires iron from both extracellular transferrin and intracellular iron pools: impact of interferon-gamma and hemochromatosis. *J Biol Chem* 277(51):49727–49734. Don't think for one second that people who have hemochromatosis are completely immune to infectious predation. There is one organism in particular that can wreak havoc on those with iron overload, *Vibrio vulnificus*. This organism is usually found in seafood and seawater, and it has a unique way of acquiring iron that makes people who have hemochromatosis highly susceptible to infection. For more on *Vibrio vulnificus* see J. J. Bullen, P. B. Spalding, C. G. Ward, and J. M. Gutteridge. 1991. Hemochromatosis, iron and septicemia caused by *Vibrio vulnificus*. *Arch Intern Med* 151(8):1606–1609. For more fun vibrio facts see the following two websites of the CDC and FDA: www.cdc.gov/ncidod/dbmd/diseaseinfo/vibriovulnificus_g.htm and www.cfsan.fda.gov/~mow/chap10.html.

14 the Vikings and hemochromatosis

For more information on the debate regarding the origins of hemochromatosis see N. Milman and P. Pedersen. 2003. Evidence that the Cys282Tyr mutation of the HFE gene originated from a population in Southern Scandinavia and spread with the Vikings. *Clin Genet* 64(1):36–47; A. Pietrangelo. 2004. Hereditary hemochromatosis—a new look at an old disease. *N Engl J Med* 350(23):2383–2397; G. Lucotte and F. Dieterlen. 2003. A European allele map of the C282Y mutation of hemochromatosis: Celtic versus Viking origin of the mutation? *Blood Cells Mol Dis* 31(2):262–267.

16 on the history of bloodletting

For a fun read on the science behind the ancient practice of bloodletting see chapter 6, "A Bloody Good Remedy," in R. S. Root-Bernstein and M. Root-Bernstein, *Honey, Mud, Maggots, and Other Medical Marvels: The Science behind Folk Remedies and Old Wives' Tales* (Boston: Houghton Mifflin, 1997); also see R. J. Weinberg, S. R. Ell, and E. D. Weinberg. 1986. Blood-letting, iron homeostasis, and human health. *Med Hypotheses* 21(4):441–443. For a paper covering a thorough history of bloodletting see G. R. Seigworth, 1980. Bloodletting over the centuries. *NY State J Med* 80(13):2022–2028. To view a surgeon's bloodletting kit from the U.S. War of Independence see http://americanhistory.si.edu/militaryhistory/exhibition/flash.html?path=1.3.r_70.

To learn more about bloodletting and fever reduction see N. W. Kasting, 1990. A rationale for centuries of therapeutic bloodletting: antipyretic therapy for febrile diseases. *Perspect Biol Med* 33(4):509–516.

19   iron, infection, Maori babies, and botulism

M. J. Murray, A. B. Murray, M. B. Murray, and C. J. Murray. 1978. The adverse effect of iron repletion on the course of certain infections. *Br Med J* 2(6145):1113–1115; R. J. Cantwell. 1972. Iron deficiency anemia of infancy: some clinical principles illustrated by the response of Maori infants to neonatal parenteral iron administration. *Clin Pediatr (Phila)* 11(8):443–449; S. S. Arnon, K. Damus, B. Thompson, et al. 1982. Protective role of human milk against sudden death from infant botulism. *J Pediatr* 100(4): 568–573.

CHAPTER II: A SPOONFUL OF SUGAR HELPS THE TEMPERATURE GO DOWN

23   How many people have diabetes?

For an updated account of the worldwide prevalence of diabetes see the World Health Organization website at www.who.int.

24   Chinese medicine

For a historical view of ancient Chinese medical practices and beliefs see J. Veith and Ti Huang, *The Yellow Emperor's Classic of Internal Medicine* (Berkeley: University of California Press, 1966); for a current overview of how ancient Chinese medical practices continue in China see V. Scheid, *Chinese Medicine in Contemporary China: Plurality and Synthesis* (Durham, NC: Duke University Press, 2002).

26   Pima Indians

For more information on the Pima Indians of the southwestern United States see diabetes.niddk.nih.gov/dm/pubs/pima/obesity/obesity.htm. For a personal account of the state of health of Pima Indians see G. P. Nabhan, *Why Some Like It Hot: Food, Genes, and Cultural Diversity* (Washington, DC: Island Press/Shearwater Books, 2004).

27 climate change and global warming

Two good books on the subject: B. M. Fagan, *The Little Ice Age: How Climate Made History, 1300–1850* (New York: Basic Books, 2000); T. F. Flannery, *The Weather Makers: How Man Is Changing the Climate and What It Means for Life on Earth* (New York: Atlantic Monthly Press, 2005).

29 Younger Dryas

S. Bondevik, J. Mangerud, H. H. Birks, et al. 2006. Changes in North Atlantic radiocarbon reservoir ages during the Allerod and Younger Dryas. *Science* 312(5779):1514–1517; National Research Council (U.S.), Committee on Abrupt Climate Change, *Abrupt Climate Change: Inevitable Surprises* (Washington, DC: National Academies Press, 2002); L. Tarasov and W. R. Peltier. 2005. Arctic freshwater forcing of the Younger Dryas cold reversal. *Nature* 435(7042):662–665; T. Correge, M. K. Gagan, J. W. Beck, et al. 2004. Interdecadal variation in the extent of South Pacific tropical waters during the Younger Dryas event. *Nature* 428(6986):927–929; C. Singer, J. Shulmeister, and B. McLea. 1998. Evidence against a significant Younger Dryas cooling event in New Zealand. *Science* 281(5378):812–814; Richard B. Alley, "Abrupt Climate Change," *Scientific American*, November 2004.

29 *Through most of the 20th century*

S. R. Weart, *The Discovery of Global Warming* (Cambridge, MA: Harvard University Press, 2003). To read about the implications of a breakdown in the Atlantic conveyer see Fred Pearce, "Faltering Currents Trigger Freeze Fear," *New Scientist*, December 3, 2005.

31 on ice cores

R. B. Alley, *The Two-Mile Time Machine: Ice Cores, Abrupt Climate Change, and Our Future* (Princeton, NJ: Princeton University Press, 2000).

33 for a peek at Europeans: before and after the Younger Dryas

C. Gamble, W. Davies, P. Pettitt, and M. Richards. 2004. Climate change and evolving human diversity in Europe during the last glacial. *Philos Trans R Soc Lond B Biol Sci* 359(1442):243–253; discussion 253–254.

34   *baseball legend Ted Williams*

Tom Verducci, "New Details Fuel Controversy Surrounding Williams' Remains," *Sports Illustrated,* August 12, 2003. If you'd like to join the "Save Ted Williams" club see www.saveted.net.

34   *Alcor Life Extension cryonics lab*

If you're interested in the latest and greatest in cryogenics at Alcor see www .alcor.org.

36   brown fat

See the following research papers for the magic of brown fat and cold tolerance: B. Cannon and J. Nedergaard. 2004. Brown adipose tissue: function and physiological significance. *Physiol Rev* 84(1):277–359; A. L. Vallerand, J. Zamecnik, and I. Jacobs. 1995. Plasma glucose turnover during cold stress in humans. *J Appl Physiol* 78(4):1296–1302; J. Watanabe, S. Kanamura, H. Tokunaga, et al. 1987. Significance of increase in glucose 6-phosphatase activity in brown adipose cells of cold-exposed and starved mice. *Anat Rec* 219(1):39–44; A. L. Vallerand, F. Perusse, and L. J. Bukowiecki. 1990. Stimulatory effects of cold exposure and cold acclimation on glucose uptake in rat peripheral tissues. *Am J Physiol* 259(5, Pt 2):R1043–R1049; A. Porras, S. Zuluaga, A. Valladares, et al. 2003. Long-term treatment with insulin induces apoptosis in brown adipocytes: role of oxidative stress. *Endocrinology* 144(12):5390–5401.

37   cold diuresis

For the controversy, history, and science behind urinating when you're cold, and the Sunderland quote, see pages 161–176 in B. M. Marriott and S. J. Carlson, Institute of Medicine (U.S.), Committee on Military Nutrition Research, *Nutritional Needs in Cold and in High-Altitude Environments: Applications for Military Personnel in Field Operations* (Washington, DC: National Academies Press, 1996).

40   the frogsicle *(Rana sylvatica)*

Elizabeth Svoboda, "Waking from a Dead Sleep," *Discover,* February 2005; K. B. Storey and J. M. Storey. 1999. Lifestyles of the cold and frozen. *The Sci-*

ences 39(3), 32–37; David A. Fahrenthold, "Looking to Frozen Frogs for Clues to Improve Human Medicine," *Seattle Times,* December 15, 2004. For more information on applications of cold tolerance to medical practice see Cold Cures, by Dr. Boris Rubinsky, at www.pbs.org/wgbh/nova/sciencenow/ 3209/05-cures.html.

44   diabetes and evolution

Sandra Blakeslee, "New Theory Places Origin of Diabetes in an Age of Icy Hardships," *New York Times,* May 17, 2005. For the original paper describing the proposed diabetes–cold tolerance link see S. Moalem, K. B. Storey, M. E. Percy, et al. 2005. The sweet thing about Type 1 diabetes: a cryoprotective evolutionary adaptation. *Med Hypotheses* 65(1):8–16. For more reading on the subject of climate change and human evolution see W. H. Calvin, *A Brain for All Seasons: Human Evolution and Abrupt Climate Change* (Chicago: University of Chicago Press, 2002).

45   fibrinogen and the cold

R. C. Hermida, C. Calvo, D. E. Ayala, et al. 2003. Seasonal variation of fibrinogen in dipper and nondipper hypertensive patients. *Circulation* 108(9):1101– 1106; V. L. Crawford, S. E. McNerlan, and R. W. Stout. 2003. Seasonal changes in platelets, fibrinogen and factor VII in elderly people. *Age Ageing* 32(6):661–665; R. W. Stout and V. Crawford. 1991. Seasonal variations in fibrinogen concentrations among elderly people. *Lancet* 338(8758):9–13.

46   *American veterans with diabetes*

This very large study tracked 285,705 American veterans for almost two years looking at blood levels of hemoglobin A1c, which is clinically used as a marker for glucose levels over an extended period of time. The hemoglobin A1c test is based on the behavior of glucose, which binds to hemoglobin irreversibly (once glucose is bound, hemoglobin is then called glycated hemoglobin or hemoglobin A1c). Since red blood cells that house hemoglobin hang on for at least two to three months before being replaced, measuring glycated hemoglobin gives clinicians and scientists a much better window into how well the diabetes is being controlled in the individual over time. For the study mentioned in the chapter see C. L. Tseng, M. Brimacombe, M. Xie, et al. 2005.

Seasonal patterns in monthly hemoglobin A1c values. *Am J Epidemiol* 161(6):565–574.

CHAPTER III: THE CHOLESTEROL ALSO RISES

50   sunlight and vitamin D

Ingfei Chen, "Sunlight, a Cancer Protector in the Guise of a Villain?" *New York Times,* August 6, 2002; M. F. Holic. 2004. Sunlight and vitamin D for bone health and prevention of autoimmune diseases, cancers, and cardiovascular disease. *Am J Clin Nutr* 80(6 Suppl):1678S–1688S; J. M. Pettifor, G. P. Moodley, F. S. Hough, et al. 1996. The effect of season and latitude on in vitro vitamin D formation by sunlight in South Africa. *S Afr Med J* 86(10):1270–1272; Anne Marie Owens, "Second-Guessing the Big Cover-up," *National Post,* February 14, 2005; V. Tangpricha, A. Turner, C. Spina, et al. 2004. Tanning is associated with optimal vitamin D status (serum 25-hydroxyvitamin D concentration) and higher bone mineral density. *Am J Clin Nutr* 80(6):1645–1649; P. T. Liu, S. Stenger, H. Li, et al. 2006. Toll-like receptor triggering of a vitamin D–mediated human antimicrobial response. *Science* 311(5768):1770–1773; A. Zitterman. 2003. Vitamin D in preventive medicine: are we ignoring the evidence? *Br J Nutr* 89(5):552–572; R. Roelandts. 2002. The history of phototherapy: something new under the sun? *J Am Acad Dermatol* 46(6):926–930.

51   seasonal variation in cholesterol levels

I. S. Ockene, D. E. Chiriboga, E. J. Stanek III, et al. 2004. Seasonal variation in serum cholesterol levels: treatment implications and possible mechanisms. *Arch Intern Med* 164(8):863–870; M. Bluher, B. Hentschel, F. Rassoul, and V. Richter. 2001. Influence of dietary intake and physical activity on annual rhythm of cholesterol concentrations. *Chronobial Int* 18(3):541–557.

52   Crohn's and suntanning

P. Koutkia, Z. Lu, T. C. Chen, and M. F. Holick. 2001. Treatment of vitamin D deficiency due to Crohn's disease with tanning bed ultraviolet B radiation. *Gastroenterology* 121(6):1485–1488.

52   folic acid and folate

L. D. Botto, A. Lisi, E. Robert-Gnansia, et al. 2005. International retrospective cohort study of neural tube defects in relation to folic acid recommendations: are the recommendations working? *BMJ* 330(7491):571; D. B. Shurtleff. 2004. Epidemiology of neural tube defects and folic acid. *Cerebrospinal Fluid Res* 1(1):5; B. Kamen. 1997. Folate and antifolate pharmacology. *Semin Oncol* 24(5 Suppl 18):S18-30–S18-39. For the report mentioned in the chapter about three mothers who gave birth to children who had neural tube defects after they had tanned during pregnancy see P. Lapunzina. 1996. Ultraviolet light–related neural tube defects? *Am J Med Genet* 67(1):106.

53   skin color

N. G. Jablonski and G. Chaplin. 2000. The evolution of human skin coloration. *J Hum Evol* 39(1):57–106; H. Y. Thong, S. H. Jee, C. C. Sun, and R. E. Boissy. 2003. The patterns of melanosome distribution in keratinocytes of human skin as one determining factor of skin colour. *Br J Dermatol* 149(3):498–505; R. L. Lamason, M. A. Mohideen, J. R. Mest, et al. 2005. SLC24A5, a putative cation exchanger, affects pigmentation in zebrafish and humans. *Science* 310(5755):1782–1786; A. J. Thody, E. M. Higgins, K. Wakamatsu, et al. 1991. Pheomelanin as well as eumelanin is present in human epidermis. *J Invest Dermatol* 97(2):340–344; Saadia Iqbal, "A New Light on Skin Color," *National Geographic Magazine*, November 2002; Nina G. Jablonski and George Chaplin, "Skin Deep," *Scientific American*, October 2002; Adrian Barnett, "Fair Enough," *New Scientist*, October 12, 2002.

54   skin cancer

For facts and figures regarding the many different types of skin cancers see the following excellent website: www.cancer.org/docroot/PED/content/ped_7_ 1_What_You_Need_To_Know_About_Skin_Cancer.asp. See also pages 57–72 in R. Ehrlich, *Nine Crazy Ideas in Science: A Few Might Even Be True* (Princeton, NJ: Princeton University Press, 2001).

55   red hair

P. Valverde, E. Healy, I. Jackson, et al. 1995. Variants of the melanocyte-stimulating hormone receptor gene are associated with red hair and fair skin in humans. *Nat Genet* 11(3):328–330; Robin L. Flanigan, "Will Rare Red-heads Be Extinct by 2100?" *Seattle Times,* May 9, 2005; T. Ha and J. L. Rees. 2001. Melanocortin 1 receptor: what's red got to do with it? *J Am Acad Dermatol* 45(6):961–964.

56   cod liver oil

See pages 10–11 in the wonderful book, R. S. Root-Bernstein and M. Root-Bernstein, *Honey, Mud, Maggots, and Other Medical Marvels: The Science Behind Folk Remedies and Old Wives' Tales* (Boston: Houghton Mifflin, 1997); K. Rajakumar. 2003. Vitamin D, cod-liver oil, sunlight, and rickets: a historical perspective. *Pediatrics* 112(2):e132–e135; M. Brustad, T. Sandanger, L. Aksnes, and E. Lund. 2004. Vitamin D status in a rural population of northern Norway with high fish liver consumption. *Public Health Nutr* 7(6):783–789; D. J. Holub and B. J. Holub. 2004. Omega-3 fatty acids from fish oils and cardiovascular disease. *Mol Cell Biochem* 263(1–2):217–225.

58   the benefits of ACHOO syndrome

H. C. Everett. 1964. Sneezing in response to light. *Neurology* 14:483–490. For more see R. Smith. 1990. Photic sneezes. *Br J Ophthalmol* 74(12):705; S. J. Peroutka and L. A. Peroutka. 1984. Autosomal dominant transmission of the "photic sneeze reflex." *N Engl J Med* 310(9):599–600; J. M. Forrester. 1985. Sneezing on exposure to bright light as an inherited response. *Hum Hered* 35(2):113–114; E. W. Benbow. 1991. Practical hazards of photic sneezing. *Br J Ophthalmol* 75(7):447.

58   Asian flush: the genetics of alcohol consumption

T. L. Wall, S. M. Horn, M. L. Johnson, et al. 2000. Hangover symptoms in Asian Americans with variations in the aldehyde dehydrogenase *(ALDH2)* gene. *J Stud Alcohol* 61(1):13–17; M. Yokoyama, A. Yokoyama, T. Yokoyama, et al. 2005. Hangover susceptibility in relation to aldehyde dehydrogenase-2 genotype, alcohol flushing, and mean corpuscular volume in Japanese workers. *Alcohol Clin Exp Res* 29(7):1165–1171; K. A. Veverka, K. L. Johnson, D. C. Mays, et al. 1997. Inhibition of aldehyde dehydrogenase by disulfiram and its metabolite methyl diethylthiocarbamoyl-sulfoxide. *Biochem Pharmacol*

53(4):511–518; Janna Chan, "Asian Flush: The Silent Killer," AsianAvenue .com, November 18, 2004.

61   the Cohan gene

See pages 140–165 in M. Z. Wahrman, *Brave New Judaism: When Science and Scripture Collide* (Hanover, NH: University Press of New England for Brandeis University Press, 2002); K. Skorecki, S. Selig, S. Blazer, et al. 1997. Y chromosomes of Jewish priests. *Nature* 385(6611):32; M. G. Thomas, K. Skorecki, H. Ben-Ami, et al. 1998. Origins of Old Testament priests. *Nature* 394(6689):138–140. This recent paper challenges the previous findings: A. Zoossmann-Diskin. 2006. Ashkenazi Levites' "Y modal haplotype" (LMH)—an artificially created phenomenon? *Homo* 57(1):87–100.

62   *"population clusters"*

For more on Dr. Henry Louis Gates see www.pbs.org/wnet/aalives/science_ dna2.html. See also Editorial. 2001. Genes, drugs and race. *Nat Genet* 29(3):239–240; Emma Daly, "DNA Tells Students They Aren't Who They Thought," *New York Times*, April 13, 2005; Marek Kohn, "This Racist Undercurrent in the Tide of Genetic Research," *Guardian*, January 17, 2006; Richard Willing, "DNA Tests to Offer Clues to Suspect's Race," *USA Today*, August 17, 2005.

64   African slave trade and hypertension

See R. Cooper and C. Rotimi. 1997. Hypertension in blacks. *Am J Hypertens* 10(7 Pt 1):804–812; M. P. Blaustein and C. E. Grim. 1991. The pathogenesis of hypertension: black-white differences. *Cardiovasc Clin* 21(3):97–114. For more information regarding "how being black affects your blood pressure" see www.mayoclinic.com/health/high-blood-pressure/HI00067. For more information on the "Middle Passage" see page 33 of N. I. Painter, *Creating Black Americans: African-American History and Its Meanings, 1619 to the Present* (New York: Oxford University Press, 2006); also see J. Postma, *The Atlantic Slave Trade* (Gainesville: University Press of Florida, 2005); Harold M. Schmeck Jr., "Study of Chimps Strongly Backs Salt's Link to High Blood Pressure," *New York Times*, October 3, 1995; Richard S. Cooper, Charles N. Rotimi, and Ryk Ward, "The Puzzle of Hypertension in African-Americans,"

Scientific American, February 1999. For specific racial/ethnic statistics see the U.S. government's Office of Minority Health website at www.omhrc.gov.

65 mortality during the Middle Passage

See pages 43–59 in J. Postma, *The Atlantic Slave Trade* (Gainesville: University Press of Florida, 2005).

68 BiDil

Stephanie Saul, "F.D.A. Approves a Heart Drug for African-Americans," *New York Times,* June 24, 2005; Kai Wright, "Death by Racism," *Dallas Morning News,* June 25, 2006; for more on this controversial drug see www.bidil. com.

68 fast versus slow drug metabolizers

I. Johansson, E. Lundqvist, L. Bertilsson, et al. 1993. Inherited amplification of an active gene in the cytochrome P450 *CYP2D* locus as a cause of ultra-rapid metabolism of debrisoquine. *Proc Natl Acad Sci U S A* 90(24):11825–11829.

69 multiple copies of genes

For more on this topic see Bob Holms, "Magic Numbers," *New Scientist,* April 8, 2006; R. Tubbs, J. Pettay, D. Hicks, et al. 2004. Novel bright field molecular morphology methods for detection of *HER2* gene amplification. *J Mol Histol* 35(6):589–594.

69 overdosing on cough syrup: the *CYP2D6* connection

Y. Gasche, Y. Daali, M. Fathi, et al. 2004. Codeine intoxication associated with ultrarapid *CYP2D6* metabolism. *N Engl J Med* 351(27):2827–2831.

69 immunity to HIV: *CCR5-Δ32*

For more on the total absence of *CCR5-Δ32* in other populations, specifically the Indian population, and the increased risk of HIV infection, see Seema Singh Bangalore, " 'Wrong' Genes May Raise AIDS Risk for Millions," *New Scientist,* April 16, 2005; Julie Clayton, "Beating the Odds," *New Scientist,* February 8, 2003; J. Novembre, A. P. Galvani, and M. Slatkin. 2005. The geographic spread of the *CCR5-Delta32* HIV-resistance allele. *PLoS Biol* 3(11): e339.

69 pharmacogenomics

T. A. Clayton, J. C. Lindon, O. Cloarec, et al. 2006. Pharmaco-metabonomic phenotyping and personalized drug treatment. *Nature* 440(7087):1073–1077; S. K. Tate and D. B. Goldstein. 2004. Will tomorrow's medicines work for everyone? *Nat Genet* 36(11 Suppl):S34–S42; I. Roots, T. Gerloff, C. Meisel, et al. 2004. Pharmacogenetics-based new therapeutic concepts. *Drug Metab Rev* 36(3–4):617–638; R. E. Cannon. 2006. A discussion of gene-environment interactions: fundamentals of ecogenetics. *Environ Health Perspect* 114(6):a382; C. G. N. Mascie-Taylor, J. Peters, and S. McGarvey, Society for the Study of Human Biology, *The Changing Face of Disease: Implications for Society* (Boca Raton, FL: CRC Press, 2004); Jo Whelan, "Where's the Smart Money Going in Biotech?" *New Scientist*, June 18, 2005; for more information see the Centers for Disease Control and Prevention website at www.cdc.gov/PCD/issues/2005/apr/04_0134.htm.

69 personalized medicine: pharmacogenomics or pharmacogenetics

For a "future model of cancer care," see the special May 26, 2006, issue of the journal *Science* 312(5777):1157–1175.

CHAPTER IV: HEY, BUD, CAN YOU DO ME A FAVA?

71 fava beans aka broad beans

See pages 40–41 in M. Toussaint-Samat, *A History of Food* (Cambridge, MA: Blackwell Reference, 1993); D. Zohary and M. Hopf, *Domestication of Plants in the Old World: The Origin and Spread of Cultivated Plants in West Asia, Europe, and the Nile Valley* (New York: Oxford University Press, 2000); J. Golenser, J. Miller, D. T. Spira, et al. 1983. Inhibitory effect of a fava bean component on the in vitro development of *Plasmodium falciparum* in normal and glucose-6-phosphate dehydrogenase deficient erythrocytes. *Blood* 61(3):507–510.

72 Pythagoras and fava beans

J. Meletis and K. Konstantopoulos. 2004. Favism—from the "avoid fava beans" of Pythagoras to the present. *Haema* 7(1):17–21.

72   *One should abstain*

Quoted in R. Parsons, "The Long History of the Mysterious Fava Bean," *Los Angeles Times,* May 29, 1996.

73   favism

K. Iwai, A. Hirono, H. Matsuoka, et al. 2001. Distribution of glucose 6-phosphate dehydrogenase mutations in Southeast Asia. *Hum Genet* 108(6):445–449; A. K. Roychoudhury and M. Nei. *Human Polymorphic Genes: World Distribution* (New York: Oxford University Press, 1988); S. H. Katz and J. Schall. 1979. Fava bean consumption and biocultural evolution. *Med Anthro* 3:459–476; S. A. Tishkoff, R. Varkonyi, N. Cahinhinan, et al. 2001. Haplotype diversity and linkage disequilibrium at human *G6PD:* recent origin of alleles that confer malarial resistance. *Science* 293(5529):455–462.

73   Korean War and favism

For more on this topic see pages 70–91 in G. P. Nabhan, *Why Some Like It Hot: Food, Genes, and Cultural Diversity* (Washington, DC: Island Press/ Shearwater Books, 2004); C. F. Ockenhouse, A. Magill, D. Smith, and W. Milhous. 2005. History of U.S. military contributions to the study of malaria. *Mil Med* 170(4 Suppl):12–16; A. S. Alving, P. E. Carson, C. L. Flanagan, and C. E. Ickes. 1956. Enzymatic deficiency in primaquine-sensitive erythrocytes. *Science* 124(3220):484–485.

74   G6PD and malaria

See pages 92–94 in E. Barnes, *Diseases and Human Evolution* (Albuquerque: University of New Mexico Press, 2005); H. Ginsburg, H. Atamna, G. Shalmiev, et al. 1996. Resistance of glucose-6-phosphate dehydrogenase deficiency to malaria: effects of fava bean hydroxypyrimidine glucosides on *Plasmodium falciparum* growth in culture and on the phagocytosis of infected cells. *Parasitology* 113(Pt 1):7–18.

75   sexy chromosomes

When it comes to sex chromosome numbers there are other possible combinations, including Turner's syndrome, which results from having only one fully functional X chromosome (X,O), or Klinefelter syndrome, where a male has an extra X chromosome (XXY).

78   "natural" poisons and protectants in the food we eat

For the original studies see A. Fugh-Berman and F. Kronenberg. 2001. Red clover *(Trifolium pratense)* for menopausal women: current state of knowledge. *Menopause* 8(5):333–337; H. W. Bennetts, E. J. Underwood, and F. L. Shier. 1946. A specific breeding problem of sheep on subterranean clover pastures in Western Australia. *Aust J Agric Res* 22:131–138; S. M. Heinonen, K. Wahala, and H. Adlercreutz. 2004. Identification of urinary metabolites of the red clover isoflavones formononetin and biochanin A in human subjects. *J Agric Food Chem* 52(22):6802–6809; M. A. Wallig, K. M. Heinz-Taheny, D. L. Epps, and T. Gossman. 2005. Synergy among phytochemicals within crucifers: does it translate into chemoprotection? *J Nutr* 135(12 Suppl): 2972S–2977S. For more information on "natural" toxins in the foods we eat see the following: K. F. Lampe, M. A. McCann, and American Medical Association, *AMA Handbook of Poisonous and Injurious Plants* (Chicago: American Medical Association, distributed by Chicago Review Press, 1985); M. Stacewicz-Sapuntzakis and P. E. Bowen. 2005. Role of lycopene and tomato products in prostate health. *Biochim Biophys Acta* 1740(2):202–205; National Research Council (U.S.), Food Protection Committee, *Toxicants Occurring Naturally in Food* (Washington, DC: National Academy of Sciences, 1973); D. R. Jacobs Jr. and L. M. Steffen. 2003. Nutrients, foods, and dietary patterns as exposures in research: a framework for food synergy. *Am J Clin Nutr* 78(3 Suppl):508S–513S; J. M. Kingsbury, *Poisonous Plants of the United States and Canada* (Englewood Cliffs, NJ: Prentice-Hall, 1964). For cassava toxicity see M. Ernesto, A. P. Cardoso, D. Nicala, et al. 2002. Persistent konzo and cyanogen toxicity from cassava in northern Mozambique. *Acta Trop* 82(3):357–362; M. L. Mlingi, M. Bokanga, F. P. Kavishe, et al. 1996. Milling reduces the goitrogenic potential of cassava. *Int J Food Sci Nutr* 47(6):445–454. For chickpea poisons see P. Smirnoff, S. Khalef, Y. Birk, and S. W. Applebaum. 1976. A trypsin and chymotrypsin inhibitor from chick peas *(Cicer arietinum)*. *Biochem J* 157(3):745–751.

79   Carl Djerassi and the Pill

For a personal account of the birth of the "Pill" see C. Djerassi, *This Man's Pill: Reflections on the 50th Birthday of the Pill* (New York: Oxford University Press,

2001) and C. Djerassi, *The Pill, Pygmy Chimps, and Degas' Horse: The Autobiography of Carl Djerassi* (New York: Basic Books, 1992).

80   Indian vetch

Leigh Dayton, "Australia Exports Poisonous 'Lentils,'" *New Scientist,* October 3, 1992; for more information see www.cropscience.org.au/icsc2004/poster/3/2/1/769_vetch.htm.

80   the deadly nightshade family

J. L. Muller. 1998. Love potions and the ointment of witches: historical aspects of the nightshade alkaloids. *J Toxicol Clin Toxicol* 36(6):617–627.

81   *Some of them ate plentifully*

R. Beverley and L. B. Wright, *The History and Present State of Virginia* (Charlottesville, VA: Dominion Books, 1968); S. Berkov, R. Zayed, and T. Doncheva. 2006. Alkaloid patterns in some varieties of *Datura stramonium. Fitoterapia* 77(3):179–182.

83   the ethnicity of capsaicin

There is considerable variation in the P450 class of enzymes in different ethnic groups, most likely the result of having lived in very different "chemical environments." This cytochrome system is what is used by the body to process or "detoxify" chemicals including prescription drugs. The following article is important in that it looks at the metabolism of capsaicin, the molecule that puts the fire in hot peppers by cytochrome P450: C. A. Reilly, W. J. Ehlhardt, D. A. Jackson, et al. 2003. Metabolism of capsaicin by cytochrome P450 produces novel dehydrogenated metabolites and decreases cytotoxicity to lung and liver cells. *Chem Res Toxicol* 16(3):336–349. These differences are also the basis of the proposed future of personalized medicine based on the genes you might have called pharmacogenomics or pharmacogenetics. See P. Gazerani and L. Arendt-Nielsen. 2005. The impact of ethnic differences in response to capsaicin-induced trigeminal sensitization. *Pain* 117(1–2):223–229.

83   capsaicin: the spice that can cause neurodegeneration
and stomach cancer

A. Mathew, P. Gangadharan, C. Varghese, and M. K. Nair. 2000. Diet and stomach cancer: a case-control study in South India. *Eur J Cancer Prev*

9(2):89–97; G. Jancso and S. N. Lawson. 1990. Transganglionic degeneration of capsaicin-sensitive C-fiber primary afferent terminals. *Neuroscience* 39(2): 501–511; D. H. Wang, W. Wu, and K. J. Lookingland. 2001. Degeneration of capsaicin-sensitive sensory nerves leads to increased salt sensitivity through enhancement of sympathoexcitatory response. *Hypertension* 37(2 Pt 2):440–443.

There are many articles regarding the benefits of capsaicin; these are just a few: E. Pospisilova and J. Palecek. 2006. Post-operative pain behavior in rats is reduced after single high-concentration capsaicin application. *Pain* [Epub June 21, 2006, ahead of print]; A. L. Mounsey, L. G. Matthew, and D. C. Slawson. 2005. Herpes zoster and postherpetic neuralgia: prevention and management. *Am Fam Physician* 72(6):1075–1080; Mary Ann Ryan, "Capsaicin Chemistry Is Hot, Hot, Hot!" *American Chemical Society,* March 24, 2003, available online at www.chemistry.org/portal/a/c/s/1/feature_ent.html?id =b90b964c5ade11d7e3d26ed9fe800100.

84   bitterness

N. Soranzo, B. Bufe, P. C. Sabeti, et al. 2005. Positive selection on a high-sensitivity allele of the human bitter-taste receptor *TAS2R16.* *Curr Biol* 15(14):1257–1265; B. Bufe, T. Hofmann, D. Krautwurst, et al. 2002. The human *TAS2R16* receptor mediates bitter taste in response to beta-glucopyranosides. *Nat Genet* 32(3):397–401.

84   supertasters

A. Drewnowski, S. A. Henderson, A. B. Shore, and A. Barratt-Fornell. 1997. Nontasters, tasters, and supertasters of *6-n-propylthiouracil* (PROP) and hedonic response to sweet. *Physiol Behav* 62(3):649–655; G. L. Goldstein, H. Daun, and B. J. Tepper. 2005. Adiposity in middle-aged women is associated with genetic taste blindness to *6-n-propylthiouracil. Obes Res* 13(6):1017–1023. See pages 118–123 in G. P. Nabhan, *Why Some Like It Hot: Food, Genes, and Cultural Diversity* (Washington, DC: Island Press/Shearwater Books, 2004).

86   potato late blight *(Phytopthora infestans)*

For a great microscopic image of potato late blight see http://helios.bto.ed .ac.uk/bto/microbes/blight.htm.

86   sensitivity to psoralen

The following paper is about a sixty-five-year-old woman who had a serious dermatological reaction after she ate a large quantity of celery root (*Apium graveolens*) and visited a tanning salon: B. Ljunggren. 1990. Severe phototoxic burn following celery ingestion. *Arch Dermatol* 126(10):1334–1336. Also see L. Wang, B. Sterling, and P. Don. 2002. Berloque dermatitis induced by "Florida water." *Cutis* 70(1):29–30; Institute of Medicine (U.S.). Committee on Identifying and Assessing Unintended Effects of Genetically Engineered Foods on Human Health, *Safety of Genetically Engineered Foods: Approaches to Assessing Unintended Health Effects* (Washington, DC: National Academies Press, 2004), 44.

87   G6PD and malaria

A. Yoshida and E. F. Roth Jr. 1987. Glucose-6-phosphate dehydrogenase of malaria parasite *Plasmodium falciparum*. *Blood* 69(5):1528–1530; C. Ruwende and A. Hill. 1998. Glucose-6-phosphate dehydrogenase deficiency and malaria. *J Mol Med* 76(8):581–588; F. P. Mockenhaupt, J. Mandelkow, H. Till, et al. 2003. Reduced prevalence of *Plasmodium falciparum* infection and of concomitant anaemia in pregnant women with heterozygous G6PD deficiency. *Trop Med Int Health* 8(2):118–124; C. Ruwende, S. C. Khoo, R. W. Snow, et al. 1995. Natural selection of hemi- and heterozygotes for G6PD deficiency in Africa by resistance to severe malaria. *Nature* 376(6537):246–249.

88   malaria

See pages 69–83 in E. Barnes, *Diseases and Human Evolution* (Albuquerque: University of New Mexico Press, 2005); and pages 715–722 of K. J. Ryan, C. G. Ray, and J. C. Sherris, *Sherris Medical Microbiology: An Introduction to Infectious Diseases* (New York: McGraw-Hill, 2004). For a wonderfully rich history of malaria see K. F. Kiple, *The Cambridge World History of Human Disease* (New York: Cambridge University Press, 1993). For an excellent account of the problems of malaria and pregnancy see the World Health Organization's website www.who.int/features/2003/04b/en/. For a review of the worldwide malarial distribution and risks for travelers including maps see www.ncid.cdc.gov/travel/yb/utils/ybGet.asp?section=dis&obj=index.htm.

88   Hippocrates

For a free online copy of Hippocrates' *On Airs, Waters, and Places* see the Massachusetts Institute of Technology (MIT) website at classics.mit.edu/ Hippocrates/airwatpl.html.

88   on miasma

M. Susser. 2001. Glossary: causality in public health science. *Epidemiol Community Health* 55:376–378.

89   air-conditioning and malaria

For a more in-depth account of this fascinating story see James Burke, "Cool Stuff," *Scientific American*, July 1997; also see chapter 10 in J. Burke, *Connections* (Boston: Little, Brown, 1995).

89   J. B. S. Haldane

J. Lederberg. 1999. J. B. S. Haldane (1949) on infectious disease and evolution. *Genetics* 153(1):1–3. For a biographical account of Haldane and his ideas see pages 141–223 in M. Kohn, *A Reason for Everything: Natural Selection and the English Imagination* (London: Faber and Faber, 2004).

91   for the beneficial chemicals in food

P. R. Mayeux, K. C. Agrawal, J. S. Tou, et al. 1988. The pharmacological effects of allicin, a constituent of garlic oil. *Agents Actions* 25(1–2):182–190; M. Zanolli. 2004. Phototherapy arsenal in the treatment of psoriasis. *Dermatol Clin* 22(4):397–406, viii; M. Heinrich and P. Bremner. 2006. Ethnobotany and ethnopharmacy—their role for anticancer drug development. *Curr Drug Targets* 7(3):239–245; X. Sun and D. D. Ku. 2006. Allicin in garlic protects against coronary endothelial dysfunction and right heart hypertrophy in pulmonary hypertensive rats. *Am J Physiol Heart Circ Physiol* (Epub May 26, 2006, ahead of print).

CHAPTER V: OF MICROBES AND MEN

95   the little dragon—Guinea worm *(Dracunculus medinensis)*

Donald G. McNeil Jr., "Dose of Tenacity Wears Down a Horrific Disease," *New York Times*, March 26, 2006. For an in-depth article on the eradication program by the Carter Center see E. Ruiz-Tiben and D. R. Hopkins. 2006. Dracunculiasis (Guinea worm disease) eradication. *Adv Parasitol* 61:275–309. For a seminal review on the topic see R. Muller. 1971. Studies on *Dracunculus medinensis* (Linnaeus). II. Effect of acidity on the infective larva. *J Helminthol* 45(2):285–288. For a real treat, such as "The parasite on attaining maturity, makes for the legs and feet," see pages 788–795 in P. Manson and P. H. Manson-Bahr, *Manson's Tropical Diseases: A Manual of the Diseases of Warm Climates* (Baltimore: W. Wood and Co. 1936). To find more information on the valiant and wide-ranging efforts of the Carter Center see www.carter center.org. For even more Guinea worm information, including how to pronounce its Latin name (dra-KUNK-you-LIE-uh-sis) properly, see the Centers for Disease Control's website www.cdc.gov/Ncidod/dpd/parasites/ dracuncu liasis/factsht_dracunculiasis.htm. Finally, for an account of Guinea worms throughout recorded history see pages 687–689 in K. F. Kiple, *The Cambridge World History of Human Disease* (New York: Cambridge University Press, 1993).

98   sexy immunity

In the chapter "dissimilar immune systems" refers to the major histocompatibility complex (MHC), so named since they were first identified and used for transplant matching. The MHC is akin to a cellular bar code that your body uses to distinguish friend from foe. The study referred to in the chapter is C. Wedekind, T. Seebeck, F. Bettens, and A. J. Paepke. 1995. MHC-dependent mate preferences in humans. *Proc Biol Sci* 260(1359):245–249; for a more friendly account of this phenomenon see Martie G. Haselton, "Love Special: How to Pick a Perfect Mate," *New Scientist,* April 29, 2006.

98   germs, germs, everywhere

F. Backhed, R. E. Ley, J. L. Sonnenburg, et al. 2005. Host-bacterial mutualism in the human intestine. *Science* 307(5717):1915–1920; S. R. Gill, M. Pop, R. T. Deboy, et al. 2006. Metagenomic analysis of the human distal gut microbiome. *Science* 312(5778):1355–1359; Rick Weiss, "Legion of Little Helpers in the Gut Keeps Us Alive," *Washington Post,* June 5, 2006; C. L. Sears. 2005. A dynamic partnership: celebrating our gut flora. *Anaerobe* 11(5):247–251; F. Guarner and J. R. Malagelada. 2003. Gut flora in health and disease. *Lancet*

361(9356):512–519; M. Heselmans, G. Reid, L. M. Akkermans, et al. 2005. Gut flora in health and disease: potential role of probiotics. *Curr Issues Intest Microbiol* 6(1):1–7; E. D. Weinberg. 1997. The *Lactobacillus* anomaly: total iron abstinence. *Perspect Biol Med* 40(4):578–583; S. Moalem, E. D. Weinberg, and M. E. Percy. 2004. Hemochromatosis and the enigma of misplaced iron: implications for infectious disease and survival. *Biometals* 17(2):135–139.

100   the orb-weaving spider *(Plesiometa argyra)*

W. G. Eberhard. 2000. Spider manipulation by a wasp larva. *Nature* 406(6793):255–256; W. G. Eberhard. 2001. Under the influence: webs and building behavior of *Plesiometa argyra (Araneae, Tetragnathidae)* when parasitized by *Hymenoepimecis argyraphaga (Hymenoptera, Ichneumonidae). Journal of Arachnology* 29:354–366; W. G. Eberhard. 2000. The natural history and behavior of *Hymenoepimecis argyraphaga (Hymenoptera, Ichneumonidae)* a parasitoid of *Plesiometa argyra (Araneae, Tetragnathidae). Journal of Hymenoptera Research* 9(2):220–240. For something less technical see Nicholas Wade, "Wasp Works Its Will on a Captive Spider," *New York Times,* July 25, 2000.

102   "The larva somehow"

The quotes are from a BBC article, "Parasite's Web of Death," July 19, 2000; for the original article see news.bbc.co.uk/2/hi/science/nature/841401 .htm.

102   the liver worm *(Dicrocoelium dentriticum)*

D. Otranto and D. Traversa. 2002. A review of dicrocoeliosis of ruminants including recent advances in the diagnosis and treatment. *Vet Parasitol* 107(4): 317–335. A website that offers a diagrammatic representation of this parasite's complex life cycle can be found at www.parasitology.informatik .uniwuerzburg.de/login/b/me14249.png.php.

103   the parasitic hairworm *(Spinochordodes tellinii)*

Shaoni Bhattacharya, "Parasites Brainwash Grasshoppers into Death Dive," *New Scientist,* August 31, 2005; the reference for the original study is D. G. Biron, L. Marche, F. Ponton, et al. 2005. Behavioural manipulation in a grasshopper harbouring hairworm: a proteomics approach. *Proc Biol Sci* 272(1577):

2117–2126; F. Thomas, A. Schmidt-Rhaesa, G. Martin, et al. 2002. Do hair-worms *(Nematomorpha)* manipulate the water seeking behaviour of their ter-restrial hosts? *J Evol Biol* 15:356–361. If you are lucky and the link is still functional, you can watch an online video of the hairworm in action as it leaves its poor drowning host at www.canal.ird.fr/canal.php?url=/prgrammes/recherches/grillons_us/index.htm.

104   the rabid bite

For all there is to know about rabies see pages 597–600 in K. J. Ryan, C. G. Ray, and J. C. Sherris, *Sherris Medical Microbiology: An Introduction to Infec-tious Diseases* (New York: McGraw-Hill, 2004).

104   *It is possible*

J. Moore. 1995. The behavior of parasitized animals—when an ant is not an ant. *Bioscience* 45:89–96. For more on parasite manipulation see J. Moore, *Par-asites and the Behavior of Animals* (New York: Oxford University Press, 2002). Other information was drawn from personal interviews with Professor Moore.

105   the feline fancier *(Toxoplasma gondii)*

For microscopic images of *T. gondii* see http://ryoko.biosci.ohio-state .edu/~parasite/toxoplasma.html. Y. Sukthana. 2006. Toxoplasmosis: beyond animals to humans. *Trends Parasitol* 22(3):137–142; E. F. Torrey and R. H. Yolken. 2003. *Toxoplasma gondii* and schizophrenia. *Emerg Infect Dis* 9(11):1375–1380; S. Bachmann, J. Schroder, C. Bottmer, et al. 2005. Psycho-pathology in first-episode schizophrenia and antibodies to *Toxoplasma gondii*. *Psychopathology* 38(2):87–90; J. P. Webster, P. H. Lamberton, C. A. Donnelly, and E. F. Torrey. 2006. Parasites as causative agents of human affective disor-ders? The impact of anti-psychotic, mood-stabilizer and anti-parasite medica-tion on *Toxoplasma gondii's* ability to alter host behaviour. *Proc Biol Sci* 273(1589):1023–1030.

109   *We found they [infected women] were more easy-going*

Quoted in Jennifer D'Angelo, "Feeling Sexy? It Could Be Your Cat," *Fox News,* November 4, 2003. See also A. Skallova, M. Novotna, P. Kolbekova, et al. 2005. Decreased level of novelty seeking in blood donors infected with *Toxoplasma*. *Neuro Endocrinol Lett* 26(5):480–486; J. Flegr, M. Preiss, J. Klose, et al. 2003. Decreased level of psychobiological factor novelty seeking and lower intelligence in men latently infected with the protozoan parasite *Toxoplasma gondii:* dopamine, a missing link between schizophrenia and toxoplasmosis? *Biol Psychol* 63(3):253–268; J. Flegr, J. Havlicek, P. Kodym, et al. 2002. Increased risk of traffic accidents in subjects with latent toxoplasmosis: a retrospective case-control study. *BMC Infect Dis* 2:11; M. Novotna, J. Hanusova, J. Klose, et al. 2005. Probable neuroimmunological link between *Toxoplasma* and cytomegalovirus infections and personality changes in the human host. *BMC Infect Dis* 5:54; R. H. Yolken, S. Bachmann, I. Ruslanova, et al. 2001. Antibodies to *Toxoplasma gondii* in individuals with first-episode schizophrenia. *Clin Infect Dis* 32(5):842–844; L. Jones-Brando, E. F. Torrey, and R. Yolken. 2003. Drugs used in the treatment of schizophrenia and bipolar disorder inhibit the replication of *Toxoplasma gondii*. *Schizophr Res* 62(3):237–244. For a few articles from the popular scientific press see James Randerson, "All in the Mind?" *New Scientist*, October 26, 2002; David Adam, "Can a Parasite Carried by Cats Change Your Personality?" *Guardian Unlimited*, September 25, 2003; *New Scientist* Editorial Staff, "Antipsychotic Drug Lessens Sick Rats' Suicidal Tendencies," *New Scientist*, January 28, 2006; Jill Neimark, "Can the Flu Bring on Psychosis?" *Discover*, October 2005.

109 why colds make us sneeze
See pages 46 and 57 in R. M. Nesse and G. C. Williams, *Why We Get Sick: The New Science of Darwinian Medicine* (New York: Times Books, 1994).

110 fancy those pinworms
To see how many children in America are infected see the CDC website www .cdc.gov/ncidod/dpd/parasites/pinworm/factsht_pinworm.htm.

111 manipulative malaria
Carl Zimmer, "Manipulative Malaria Parasite Makes You More Attractive (to Mosquitoes)," *New York Times*, August 9, 2005.

112 pediatric autoimmune neuropsychiatric disorder associated with streptococcal infection (PANDAS)

S. E. Swedo, H. L. Leonard, M. Garvey, et al. 1998. Pediatric autoimmune neuropsychiatric disorders associated with streptococcal infections: clinical description of the first 50 cases. *Am J Psychiatry* 155(2):264–271; L. A. Snider and S. E. Swedo. 2004. PANDAS: current status and directions for research. *Mol Psychiatry* 9(10):900–907; R. C. Dale, I. Heyman, G. Giovannoni, and A. W. Church. 2005. Incidence of anti-brain antibodies in children with obsessive-compulsive disorder. *Br J Psychiatry* 187:314–319; S. E. Swedo and P. J. Grant. 2005. Annotation: PANDAS: a model for human autoimmune disease. *J Child Psychol Psychiatry* 46(3):227–234; C. Heubi and S. R. Shott. 2003. PANDAS: pediatric autoimmune neuropsychiatric disorders associated with streptococcal infections—an uncommon, but important indication for tonsillectomy. *Int J Pediatr Otorhinolaryngol* 67(8):837–840; Anahad O'Connor, "Can Strep Bring On an Anxiety Disorder?" *New York Times,* December 14, 2005; Lisa Belkin, "Can You Catch Obsessive-Compulsive Disorder?" *New York Times,* May 22, 2005; Nicholas Bakalar, "Tonsil-Adenoid Surgery May Help Behavior, Too," *New York Times,* April 4, 2006.

114 *It is intriguing to speculate*

Quoted on page 205 in N. E. Beckage, *Parasites and Pathogens: Effects on Host Hormones and Behavior* (New York: Chapman & Hall, 1997).

115 sick and lonely spiny lobsters

D. C. Behringer, M. J. Butler, and J. D. Shields. 2006. Ecology: avoidance of disease by social lobsters. *Nature* 441(7092):421.

116 why we fear strangers

J. Faulkner, M. Schaller, J. H. Park, and L. A. Duncan. 2004. Evolved disease-avoidance mechanisms and contemporary xenophobic attitudes. *Group Processes & Intergroup Relations* 4:333–353; L. Rózsa. 2000. Spite, xenophobia, and collaboration between hosts and parasites. *Oikos* 91:396–400; R. Kurzban and M. R. Leary. 2001. Evolutionary origins of stigmatization: the functions of social exclusion. *Psychol Bull* 127(2):187–208.

116 " 'Superbugs' Spread Fear"

Anita Manning, " 'Superbugs' Spread Fear Far and Wide," *USA Today*, May 10, 2006; "Rising Deadly Infections Puzzle Experts," Associated Press, May 12, 2006; Abigail Zuger, "Bacteria Run Wild, Defying Antibiotics," *New York Times*, March 2, 2004.

117 staph is back

For all the facts about staph see K. J. Ryan, C. G. Ray, and J. C. Sherris, *Sherris Medical Microbiology: An Introduction to Infectious Diseases* (New York: Mc-Graw-Hill, 2004). For the discovery of penicillin see page 216 in T. Rosebury, *Microbes and Morals: The Strange Story of Venereal Disease* (New York: Viking Press, 1971); M. C. Enright, D. A. Robinson, G. Randle, et al. 2002. The evolutionary history of methicillin-resistant *Staphylococcus aureus* (MRSA). *Proc Natl Acad Sci* 99(11):7687–7692; L. B. Rice. 2006. Antimicrobial resistance in gram-positive bacteria. *Am J Med* 119(6 Suppl 1):S11–S19, discussion S62–S70; K. Hiramatsu, H. Hanaki, T. Ino, et al. 1997. Methicillin-resistant *Staphylococcus aureus* clinical strain with reduced vancomycin susceptibility. *J Antimicrob Chemother* 40(1):135–136; Allison George, "March of the Super Bugs," *New Scientist*, July 19, 2003.

121 We should be taking control

For the quotes from Ewald, and more on the evolution of the pathogenicity of cholera, see the PBS website www.pbs.org/wgbh/evolution/library/01/6/text_pop/1_016_06.html. Roger Lewin, "Shock of the Past for Modern Medicine: A Radical Approach to Medicine Seeks to Explain Diseases and Their Symptoms as a Legacy of Our Evolution: Can Darwinism Lead to Better Treatments?" *New Scientist*, October 23, 1993; P. W. Ewald, *Evolution of Infectious Disease* (New York: Oxford University Press, 1994); P. W. Ewald. 2004. Evolution of virulence. *Infect Dis Clin North Am* 18(1): 1–15; Paul Ewald, "The Evolution of Virulence," *Scientific American*, April 1993. For an interesting interview with Professor Ewald that's available online see www.findarticles.com/p/articles/mi_m1430/is_n6_v17/ai_16595653. Another good article is available online at www.cdc.gov/ncidod/eid/vol2no4/ewald.htm.

CHAPTER VI: JUMP INTO THE GENE POOL

125   Edward Jenner and smallpox

A. J. Stewart and P. M. Devlin. 2006. The history of the smallpox vaccine. *J Infect* 52(5):329–334; for more on the history of smallpox see K. F. Kiple, *The Cambridge World History of Human Disease* (New York: Cambridge University Press, 1993).

127   how many genes do we have?

L. D. Stein. 2004. Human genome: end of the beginning. *Nature* 431(7011): 915–916.

128   junk DNA

"The word: Junk DNA," *New Scientist*, November 19, 2005; Wayt Gibbs, "The Unseen Genome: Gems among the Junk," *Scientific American*, November 2003. This article is a little dated but still good: Natalie Angier, "Keys Emerge to Mystery of 'Junk' DNA," *New York Times*, June 28, 1994. Junk DNA finally gets an upgrade, in P. Andolfatto. 2005. Adaptive evolution of non-coding DNA in *Drosophila*. *Nature* 437(7062):1149–1152; James Kingsland, "Wonderful Spam," *New Scientist*, May 29, 2004.

129   mitochondria

For more on the fascinating story behind mitochondria, those lovable little organelles, see Philip Cohen, "The Force," *New Scientist*, February 26, 2000.

130   solar radiation

D. S. Smith, J. Scalo, and J. C. Wheeler. 2004. Importance of biologically active aurora-like ultraviolet emission: stochastic irradiation of Earth and Mars by flares and explosions. *Orig Life Evol Biosph* 34(5):513–532; K. G. McCracken, J. Beer, and F. B. McDonald. 2004. Variations in the cosmic radiation, 1890–1986, and the solar and terrestrial implications. *Ad Space Res* 34:397–406; T. I. Pulkkinen, H. Nevanlinna, P. J. Pulkkinen, and M. Lockwood. 2001. The Sun-Earth connection in time scales from years to decades and centuries. *Space Science Reviews* 95(1/2):625–637; H. S. Hudson, S. Silva, and M. Woodard. 1982. The effects of sunspots on solar radiation. *Solar Physics* 76:211–219; Malcolm W. Browne, "Flu Time: When the Sunspots Are

Jumping?" *New York Times,* January 25, 1990; F. Hoyle and N. C. Wickrama-singhe. 1990. Sunspots and influenza. *Nature* 343(6256):304; J. W. Yeung. 2006. A hypothesis: sunspot cycles may detect pandemic influenza A in 1700–2000 A.D. *Med Hypotheses* 67(5):1016–1022.

133   shuffling genes

One example of the incredible power of genes to rearrange themselves is seen in the fruit fly gene called *Dscam.* Genes are rearranged through an enzymatic "card shuffler" called a spliceosome. The *Dscam* gene is truly amazing, since it can produce 38,016 different and distinct proteins. For a sample of papers on *Dscam* see J. M. Kreahling and B. R. Graveley. 2005. The iStem, a long-range RNA secondary structure element required for efficient exon inclusion in the *Drosophila Dscam* pre-mRNA. *Mol Cell Biol* 25(23):10251–10260; A. M. Ce-lotto and B. R. Graveley. 2001. Alternative splicing of the *Drosophila Dscam* pre-mRNA is both temporally and spatially regulated. *Genetics* 159(2):599–608; G. Parra, A. Reymond, N. Dabbouseh, et al. 2006. Tandem chimerism as a means to increase protein complexity in the human genome. *Genome Res* 16(1):37–44.

133   biological Kaizen

A book on the ethos of continual improvement from a business perspective: M. Ima, *Kaizen (Ky'zen), the Key to Japan's Competitive Success* (New York: Random House Business Division, 1986).

133   the surprise "KO": animals with missing genes are normal

See pages 64–82 in M. Morange, *The Misunderstood Gene* (Cambridge, MA: Harvard University Press, 2001).

134   life at the end of the genome

See pages 183–198 in I. Moss, *What Genes Can't Do* (Cambridge, MA: MIT Press, 2003); H. Pearson. 2006. Genetics: what is a gene? *Nature* 441(7092): 398–401.

134   Jean-Baptiste Lamarck

A very strange period in the study of heritability by Soviet scientists was shaped largely by Trofim Denisovich Lysenko. Lysenkoism, as it came to be known, was an extreme twist on acquired characteristics. For more on this fascinating period in history see pages 183–187 in M. Kohn, *A Reason for Everything: Natural Selection and the English Imagination* (London: Faber and Faber, 2004); C. Darwin, *The Origin of the Species* (New York: Fine Creative Media, 2003).

137 *attention undoubtedly will be centered*

"The Significance of Responses of the Genome Challenge," December 8, 1983, available at www.nobelprize.org/nobel_prizes/medicine/laureates/1983/mcclintock-lecture.pdf. For a great online resource from the U.S. National Library of Medicine about McClintock see www.profiles.nlm.nih.gov/LL/Views/Exhibit/narrative/biographical.html. Also see Vidyanand Nanjundiah, "Barbara McClintock and the Discovery of Jumping Genes," *Resonance*, October 1996.

138 a fly named "Methuselah"

Y. J. Lin, L. Seroude, and S. Benzer. 1998. Extended life-span and stress resistance in the *Drosophila* mutant methuselah. *Science* 282(5390):943–946; for a fun article on how you might be the next Methuselah see Kate Douglas, "How to Live to 100 . . . and Enjoy It," *New Scientist*, June 3, 2006.

139 stars of the transposable world: *gypsy, mtanga, Castaway, Evelknievel, mariner*

J. Modolell, W. Bender, and M. Meselson. 1983. *Drosophila melanogaster* mutations suppressible by the suppressor of Hairy-wing are insertions of a 7.3-kilobase mobile element. *Proc Natl Acad Sci U S A* 80(6):1678–1682; C. J. Rohr, H. Ranson, X. Wang, and N. J. Besansky. 2002. Structure and evolution of mtanga, a retrotransposon actively expressed on the Y chromosome of the African malaria vector *Anopheles gambiae. Mol Biol Evol* 19(2):149–162; T. E. Bureau, P. C. Ronald, and S. R. Wessler. 1996. A computer-based systematic survey reveals the predominance of small inverted-repeat elements in wild-type rice genes. *Proc Natl Acad Sci U S A* 93(16):8524–8529; S. Henikoff and L. Comai. 1998. A DNA methyltransferase homolog with a chromodomain exists in multiple polymorphic forms in *Arabidopsis. Genetics* 149(1):

307–318; J. W. Jacobson, M. M. Medhora, and D. L. Hartl. 1986. Molecular structure of a somatically unstable transposable element in *Drosophila*. *Proc Natl Acad Sci U S A* 83(22):8684–8688; S. M. Miller, R. Schmitt, and D. L. Kirk. 1993. Jordan, an active Volvox transposable element similar to higher plant transposons. *Plant Cell* 5(9):1125–1138.

140   *The genome has long been thought*

G. G. Dimijian. 2000. Pathogens and parasites: strategies and challenges. *Proc (Bayl Univ Med Cent)* 13(1):19–29.

141   **making a mutant**

The papers mentioned in the study are J. Cairns, J. Overbaugh, and S. Miller. 1988. The origin of mutants. *Nature* 335(6186):142–145; B. G. Hall. 1990. Spontaneous point mutations that occur more often when advantageous than when neutral. *Genetics* 126(1):5–16; S. M. Rosenberg. 1997. Mutation for survival. *Curr Opin Genet Dev* 7(6):829–834; J. Torkelson, R. S. Harris, M. J. Lombardo, et al. 1997. Genome-wide hypermutation in a subpopulation of stationary-phase cells underlies recombination-dependent adaptive mutation. *Embo J* 16(11):3303–3311; P. L. Foster. 1997. Nonadaptive mutations occur on the F' episome during adaptive mutation conditions in *Escherichia coli*. *J Bacteriol* 179(5):1550–1554; O. Tenaillon, E. Denamur, and I. Matic. 2004. Evolutionary significance of stress-induced mutagenesis in bacteria. *Trends Microbiol* 12(6):264–270. For Matic's study that was mentioned in the chapter see I. Bjedov, O. Tenaillon, B. Gerard, et al. 2003. Stress-induced mutagenesis in bacteria. *Science* 300(5624):1404–1409.

144   **genes that increase the risk of cancer: *BRCA1* and *BRCA2***

For a good resource that covers many common conditions and their associated genetics see P. Reilly, *Is It in Your Genes? The Influence of Genes on Common Disorders and Diseases That Affect You and Your Family* (Cold Spring Harbor, NY: Cold Spring Harbor Laboratory Press, 2004).

144   *"You wouldn't want that added element"*

Professor Fred Gage, quoted in a press release available at genome.wellcome .ac.uk/doc_WTD020792.html. For the jumping genes within the brain article see A. R. Muotri, V. T. Chu, M. C. Marchetto, et al. 2005. Somatic mosaicism in neuronal precursor cells mediated by LI retrotransposition. *Nature* 435(7044):903–910.

146 *Hermes behaves more like*

Nancy Craig, quoted in a press release available at www.hopkinsmedicine .org/Press_releases/2004/12_23_04.html. The study Nancy Craig is talking of is L. Zhou, R. Mitra, P. W. Atkinson, et al. 2004. Transposition of hAT elements links transposable elements and V(D)J recombination. *Nature* 432(7020):995–1001. Also see M. Bogue and D. B. Roth. 1996. Mechanism of V(D)J recombination. *Curr Opin Immunol* 8(2):175–180.

146 **our genes show signs of jumping genes**

For more on this fascinating idea see James Kingsland, "Wonderful Spam," *New Scientist*, May 29, 2004.

146 *have been remodeling*

Jef Boeke, quoted at www.eurekalert.org/pub_releases/2002-08/jhmigc081502 .php. The original paper that Professor Boeke is commenting on is D. E. Symer, C. Connelly, S. T. Szak, et al. 2002. Human 11 retrotransposition is associated with genetic instability in vivo. *Cell* 110(3):327–338.

148 **half the genome from jumping genes**

P. Medstrand, L. N. van de Lagemaat, C. A. Dunn, et al. 2005. Impact of transposable elements on the evolution of mammalian gene regulation. *Cytogenet Genome Res* 110(1–4):342–352; W. Makalowski. 2001. The human genome structure and organization. *Acta Biochim Pol* 48(3):587–598.

150 *at least 8 percent of the human genome is composed of retroviruses*

J. F. Hughes and J. M. Coffin. 2004. Human endogenous retrovirus K solo-LTR formation and insertional polymorphisms: implications for human and viral evolution. *Proc Natl Acad Sci U S A* 101(6):1668–1672; S. Mi, X. Lee, X. Li, et al. 2000. Syncytin is a captive retroviral envelope protein involved in human placental morphogenesis. *Nature* 403(6771):785–789; J. P. Moles, A. Tesniere, and J. J. Guilhou. 2005. A new endogenous retroviral sequence is expressed in skin of patients with psoriasis. *Br J Dermatol* 153(1):83–89.

151 *"the successful genetic patterns"*

See pages 1–10 in Salvador E. Luria, *Virus Growth and Variation,* B. Lacey and I. Isaacs, eds. (Cambridge: Cambridge University Press, 1959).

151 *There's a very strong*

Luis Villarreal, personal communication. For more of his work see L. P. Villarreal. 2004. Can viruses make us human? *Proc Am Phil So* 148(3):296–323. L. P. Villarreal, *Viruses and the Evolution of Life* (Washington, DC: ASM Press, 2005); L. P. Villareal. 1997. On viruses, sex, and motherhood. *J Virol* 71(2):859–865.

151 **more on viruses as a precursor to life and evolution**

Charles Siebert, "Unintelligent Design," *Discover,* March 2006; M. Syvanen. 1984. The evolutionary implications of mobile genetic elements. *Annu Rev Genet* 18:271–293; D. J. Hedges and M. A. Batzer. 2005. From the margins of the genome: mobile elements shape primate evolution. *Bioessays* 27(8):785–794; M. G. Kidwell and D. R. Lisch. 2001. Perspective: transposable elements, parasitic DNA, and genome evolution. *Evolution Int J Org Evolution* 55(1): 1–24; J. Brosius. 2005. Echoes from the past—are we still in an RNP world? *Cytogenet Genome Res* 110(1–4):8–24; C. Biemont and C. Vieira. 2005. What transposable elements tell us about genome organization and evolution: the case of *Drosophila. Cytogenet Genome Res* 110(1–4):25–34; P. Medstrand, L. N. van de Lagemaat, C. A. Dunn, et al. 2005. Impact of transposable elements on the evolution of mammalian gene regulation. *Cytogenet Genome Res* 110(1–4):342–352.

CHAPTER VII: METHYL MADNESS: ROAD TO THE FINAL PHENOTYPE

155 *American children are overweight or obese*

For a book on the topic see F. M. Berg, *Underage & Overweight: The Childhood Obesity Crisis: What Every Family Needs to Know* (Long Island City, NY: Hatherleigh Press, 2005). On the marketing of fast food to children, for younger readers and adults alike, see E. Schlosser and C. Wilson, *Chew on This: The*

Unhappy Truth about Fast Food (Boston: Houghton Mifflin Company, 2006). For a short online article from the University of California including useful references see news.ucanr.org/mediakits/Nutrition/nutritionfactsheet.shtml. For an in-depth account of "Behavioral Risk Factor Surveillance" from the CDC see www.cdc.gov/mmwr/preview/mmwrhtml/ss4906a1.htm. Also see W. H. Dietz and T. N. Robinson. 2005. Clinical practice: overweight children and adolescents. *N Engl J Med* 352(20):2100–2109; D. S. Freedman, W. H. Dietz, S. R. Srinivasan, and G. S. Berenson. 1999. The relation of overweight to cardiovascular risk factors among children and adolescents: the Bogalusa Heart Study. *Pediatrics* 103(6 Pt 1):1175–1182. S. J. Olshansky, D. J. Passaro, R. C. Hershow, et al. 2005. A potential decline in life expectancy in the United States in the 21st century. *N Engl J Med* 352(11):1138–1145; Philip Cohen, "You Are What Your Mother Ate, Suggests Study," *New Scientist*, August 4, 2003. The study that the *New Scientist* article is referring to is R. A. Waterland and R. L. Jirtle. 2003. Transposable elements: targets for early nutritional effects on epigenetic gene regulation. *Mol Cell Biol* 23(15):5293–5300; Alison Motluk, "Life Sentence," *New Scientist*, October 30, 2004.

156   flicking the switch: environmental factors

Rowan Hooper, "Mendel's Laws of Inheritance Challenged," *New Scientist*, May 27, 2006; Rowan Hooper, "Men Inherit Hidden Cost of Dad's Vices," *New Scientist*, January 6, 2006; E. Jablonka and M. J. Lamb, *Evolution in Four Dimensions: Genetic, Epigenetic, Behavioral, and Symbolic Variation in the History of Life* (Cambridge, MA: MIT Press, 2005); R. A. Waterland and R. L. Jirtle. 2003. Transposable elements: targets for early nutritional effects on epigenetic gene regulation. *Mol Cell Biol* 23(15):5293–5300; Gaia Vince, "Pregnant Smokers Increase Grandkids' Asthma Risk," *New Scientist*, April 11, 2005; Q. Li, S. Guo-Ross, D. V. Lewis, et al. 2004. Dietary prenatal choline supplementation alters postnatal hippocampal structure and function. *J Neurophysiol* 91(4):1545–1555; Shaoni Bhattacharya, "Nutrient During Pregnancy Super-Charges' Brain," *New Scientist*, March 12, 2004; Leslie A. Pray, "Dieting for the Genome Generation," *The Scientist*, January 17, 2005; Anne Underwood and Jerry Adler, "Diet and Genes," *Newsweek*, January 24, 2005.

159   *We have long known*

Randy Jirtle, quoted in a press release, Duke University Medical Center, available at www.dukemednews.org/news/article.php?id=6804. For the full article see R. A. Waterland and R. L. Jirtle. 2003. Transposable elements: targets for early nutritional effects on epigenetic gene regulation. *Mol Cell Biol* 23(15):5293–5300; Leslie A. Pray, "Epigenetics: Genome, Meet Your Environment: As the Evidence Accumulates for Epigenetics, Researchers Reacquire a Taste for Lamarkism," *The Scientist,* July 5, 2004; I. C. Weaver, N. Cervoni, F. A. Champagne, et al. 2004. Epigenetic programming by maternal behavior. *Nat Neurosci* 7(8):847–854; E. W. Fish, D. Shahrokh, R. Bagot, et al. 2004. Epigenetic programming of stress responses through variations in maternal care. *Ann N Y Acad Sci* 1036:167–180; A. D. Riggs and Z. Xiong. 2004. Methylation and epigenetic fidelity. *Proc Natl Acad Sci U S A* 101(1): 4–5.

160   baby voles

C. R. Camargo, E. Colares, and A. M. Castrucci. 2006. Seasonal pelage color change: news based on a South American rodent. *An Acad Bras Cienc* 78(1):77–86.

161   *Daphnia* gives birth to babies with helmets

J. L. Brooks. 1965. Predation and relative helmet size in cyclomorphic *Daphnia. Proc Natl Acad Sci U S A* 53(1):119–126; J. Pijanowska and M. Kloc. 2004. *Daphnia* response to predation threat involves heat-shock proteins and the actin and tubulin cytoskeleton. *Genesis* 38(2):81–86.

161   insects change their colors

M. Enserink. 2004. Entomology: an insect's extreme makeover. *Science* 306(5703):1881.

161   what mama lizard smelled

R. Richard Shine and S. J. Downes. 1999. Can pregnant lizards adjust their offspring phenotypes to environmental conditions? *Oecologia* 119(1):1–8.

162   maternal effect

P. D. Gluckman and M. Hanson, *The Fetal Matrix: Evolution, Development, and Disease* (New York: Cambridge University Press, 2005).

163   Barker Hypothesis

Shaoni Bhattacharya, "Fattening Up Skinny Toddlers Risks Heart Health," *New Scientist,* October 27, 2005; C. N. Hales and D. J. Barker. 2001. The thrifty phenotype hypothesis. *Br Med Bull* 60:5–20.

164   *the* first four days *of pregnancy*

W. Y. Kwong, A. E. Wild, P. Roberts, et al. 2000. Maternal undernutrition during the preimplantation period of rat development causes blastocyst abnormalities and programming of postnatal hypertension. *Development* 127(19):4195–4202. For a good review of the topic see V. M. Vehaskari and L. L. Woods. 2005. Prenatal programming of hypertension: lessons from experimental models. *J Am Soc Nephrol* 16(9):2545–2556.

164   men who smoke before puberty

Rowan Hooper, "Men Inherit Hidden Cost of Dad's Vices," *New Scientist,* January 6, 2006; M. E. Pembrey, L. O. Bygren, G. Kaati, et al. 2006. Sex-specific, male-line transgenerational responses in humans. *Eur J Hum Genet* 14(2):159–166. The quote from Marcus Pembrey is from E. Pennisi. 2005. Food, tobacco, and future generations. *Science* 310(5755):1760–1761.

166   grandmothers who smoked while pregnant

Gaia Vince, "Pregnant Smokers Increase Grandkids' Asthma Risk," *New Scientist,* April 11, 2005.

166   "Hunger Winter"

L. H. Lumey, A. C. Ravelli, L. G. Wiessing, et al. 1993. The Dutch Famine Birth Cohort Study: design, validation of exposure, and selected characteristics of subjects after 43 years follow-up. *Paediatr Perinat Epidemiol* 7(4):354–367; A. D. Stein, A. C. Ravelli, and L. H. Lumey. 1995. Famine, third-trimester pregnancy weight gain, and intrauterine growth: the Dutch Famine Birth

Cohort Study. *Hum Biol* 67(1):135–150; L. H. Lumey, A. D. Stein, and A. C. Ravelli. 1995. Timing of prenatal starvation in women and birth weight in their first and second born offspring: the Dutch Famine Birth Cohort Study. *Eur J Obstet Gynecol Reprod Biol* 61(1):23–30; L. H. Lumey and A. D. Stein. 1997. In utero exposure to famine and subsequent fertility: the Dutch Famine Birth Cohort Study. *Am J Public Health* 87(12):1962–1966; A. D. Stein and L. H. Lumey. 2000. The relationship between maternal and offspring birth weights after maternal prenatal famine exposure: the Dutch Famine Birth Cohort Study. *Hum Biol* 72(4):641–654.

167   *Our findings show*

R. A. Waterland and R. L. Jirtle. 2003. Transposable elements: targets for early nutritional effects on epigenetic gene regulation. *Mol Cell Biol* 23(15):5293–5300.

172   *We believe these different epigenetic patterns*

Christen Brownlee, "Nurture Takes the Spotlight," *Science News*, June 24, 2006.

172   **Epigenomics**

The company website can be found at www.epigenomics.de/en/Company/. For more on the subject of epigenetics see G. Riddihough and E. Pennisi. 2001. The evolution of epigenetics. *Science* 293(5532):1063; E. Jablonka and M. J. Lamb. 2002. The changing concept of epigenetics. *Ann N Y Acad Sci* 981:82–96; V. K. Rakyan, J. Preis, H. D. Morgan, and E. Whitelaw. 2001. The marks, mechanisms and memory of epigenetic states in mammals. *Biochem J* 356(Pt 1):1–10.

172   **smoking and methylation**

D. H. Kim, H. H. Nelson, J. K. Wiencke, et al. 2001. p16(INK4a) and histology-specific methylation of *CpG* islands by exposure to tobacco smoke in non–small cell lung cancer. *Cancer Res* 61(8):3419–3424; H. Enokida, H. Shiina, S. Urakami, et al. 2006. Smoking influences aberrant *CpG* hypermethylation of multiple genes in human prostate carcinoma. *Cancer* 106(1): 79–86.

173 *"We'd like to use the degree"*

Dr. Dhananjaya Saranath, quoted at www.telegraphindia.com/1050214/asp/
knowhow/story_4376851.asp.

174 folic acid and neural tube defects

There is a very large literature on this subject; for a sample paper (albeit a bit
dated but still good) see MRC Vitamin Study Research Group. 1991. Preven-
tion of neural tube defects: results of the Medical Research Council Vitamin
Study. *Lancet* 338(8760):131–137, and more to the point: C. M. Ulrich and
J. D. Potter. 2006. Folate supplementation: too much of a good thing? *Cancer
Epidemiol Biomarkers Prev* 15(2):189–193.

175 betamethasone and hyperactivity

For the University of Toronto study mentioned in the chapter see A. Kapoor,
E. Dunn, A. Kostaki, et al. 2006. Fetal programming of hypothalamo-
pituitary-adrenal function: prenatal stress and glucocorticoids. *J Physiol*
572(Pt 1):31–44; P. Erdeljan, M. H. Andrews, J. F. MacDonald, and
S. G. Matthews. 2005. Glucocorticoids and serotonin alter glucocorticoid receptor mRNA levels in fetal guinea-pig hippocampal neurons, in vitro. *Reprod
Fertil Dev* 17(7):743–749. The quote in the chapter, "terrifying beyond comprehension," is from Alison Motluk, "Pregnancy Drug Can Affect Grandkids
Too," *New Scientist*, December 3, 2005.

176 *This is the first approved drug*

Peter Jones, quoted in Lori Oliwenstein, "USC Cancer Researchers Examine
Potential of Epigenetics in Nature," *HSC Weekly*, May 28, 2004.

176 *It is apparent*

G. Egger, G. Liang, A. Aparicio, and P. A. Jones. 2004. Epigenetics in human
disease and prospects for epigenetic therapy. *Nature* 429(6990):457–463.

176 Johns Hopkins and azacitidine

D. Gius, H. Cui, C. M. Bradbury, et al. 2004. Distinct effects on gene expression of chemical and genetic manipulation of the cancer epigenome revealed

by a multimodality approach. *Cancer Cell* 6(4):361–371; R. S. Tuma. 2004. Silencing the critics: studies move closer to answering epigenetic questions. *J Natl Cancer Inst* 96(22):1652–1653; M. Z. Fang, Y. Wang, N. Ai, et al. 2003. Tea polyphenol (-)-epigallocatechin-3-gallate inhibits DNA methyltransferase and reactivates methylation-silenced genes in cancer cell lines. *Cancer Res* 63(22):7563–7570.

177   *What is good in small amounts*
Dana Dolinoy, quoted in a press release available at www.dukemednews.org/news/article.php?id=9584. D. C. Dolinoy, J. R. Weidman, R. A. Waterland, and R. L. Jirtle. 2006. Maternal genistein alters coat color and protects Avy mouse offspring from obesity by modifying the fetal epigenome. *Environ Health Perspect* 114(4):567–572. Also see M. Z. Fang, D. Chen, Y. Sun, et al. 2005. Reversal of hypermethylation and reactivation of *p16INK4a*, *RARbeta*, and *MGMT* genes by genistein and other isoflavones from soy. *Clin Cancer Res* 11(19 Pt 1):7033–7041.

178   pregnancy and stress
For the study about 9/11 see R. Catalano, T. Bruckner, J. Gould, et al. 2005. Sex ratios in California following the terrorist attacks of September 11, 2001. *Hum Reprod* 20(5):1221–1227; for the study mentioned about the stress faced by East German mothers during reunification see R. A. Catalano. 2003. Sex ratios in the two Germanies: a test of the economic stress hypothesis. *Hum Reprod* 18(9):1972–1975; for the study after the war in Slovenia see B. Zorn, V. Sucur, J. Stare, and H. Meden-Vrtovec. 2002. Decline in sex ratio at birth after 10-day war in Slovenia: brief communication. *Hum Reprod* 17(12):3173–3177; for the effects of the Kobe earthquake on birth ratios see M. Fukuda, K. Fukuda, T. Shimizu, and H. Moller. 1998. Decline in sex ratio at birth after Kobe earthquake. *Hum Reprod* 13(8):2321–2322; Hazel Muir, "Women Who Believe in Long Life Bear Sons," *New Scientist*, August 4, 2004; for the original study see S. E. Johns. 2004. Subjective life expectancy predicts offspring sex in a contemporary British population. *Proc Biol Sci* 271(Suppl 6):S474–S476; Will Knight, "9/11 Babies Inherit Stress from Mothers," *New Scientist*, May 3, 2005.

180   *"all the pages of a manual"*

National Human Genome Research Institute, www.genome.gov/11006943.

181   *the Human Epigenome Project*

Shaoni Bhattacharya, "Human Gene On/Off Switches to Be Mapped," *New Scientist,* October 7, 2003; P. A. Jones and R. Martienssen. 2005. A blueprint for a Human Epigenome Project: the AACR Human Epigenome Workshop. *Cancer Res* 65(24):11241–11246. A brief online article can be found from the American Association for Cancer Research at www.aacr.org/Default .aspx?p=6336&d=562.

CHAPTER VIII: THAT'S LIFE: WHY YOU AND YOUR iPOD
MUST DIE

183   Seth Cook

Carol Smith, "Lessons from a Boy Growing Old before His Time," *Seattle Post-Intelligencer Reporter,* September 16, 2004; there's also an ABC News story about Seth at abcnews.go.com/GMA/Health/story?id=1445002. For more information on this condition visit the Hutchinson-Gilford Progeria Syndrome Network website at www.hgps.net; also the Progeria Research Foundation has an excellent website with a lot of information at www .progeriaresearch.org/progeria_101.html.

184   researchers announce finding mutation that causes progeria

M. Eriksson, W. T Brown, L. B. Gordon, et al. 2003. Recurrent de novo point mutations in lamin A cause Hutchinson-Gilford progeria syndrome. *Nature* 423(6937):293–298.

185   *reported in* Science

P. Scaffidi and T. Misteli. 2006. Lamin A–dependent nuclear defects in human aging. *Science* 312(5776):1059–1063.

185   Leonard Hayflick and his number

L. Hayflick. 1965. The limited in vitro lifetime of human diploid cell strains. *Exp Cell Res* 37:614–616; D. Josefson. 1998. US scientists extend the life of human cells. *BMJ* 316:247–252; L. Hayflick. 2000. The illusion of cell immortality. *Br J Cancer* 83(7):841–846.

186   cancer versus other conditions

See Cancer Facts and Figures—2006 from the American Cancer Society's website at www.cancer.org/downloads/STT/CAFF2006PWSecured.pdf. Also see T. Thom, N. Haase, W. Rosamond, et al. 2006. Heart disease and stroke statistics—2006 update: a report from the American Heart Association Statistics Committee and Stroke Statistics Subcommittee. *Circulation* 113(6):e85–151.

187   most cancer cells use telomerase

See an online article from the Whitehead Institute's website at www.wi.mit .edu/news/archives/1997/rw_0814.html.

188   stem cells

There is a bounty of articles on this topic; for one that is a little dated but well written see Nicholas Wade, "Experts See Immortality in Endlessly Dividing Cells," *New York Times*, November 17, 1998.

189   long life and DNA repair

G. A. Cortopassi and E. Wang. 1996. There is substantial agreement among interspecies estimates of DNA repair activity. *Mech Ageing Dev* 91(3): 211–218.

191   biogenic obsolescence

For a fascinating account of "planned obsolescence" see G. Slade, *Made to Break: Technology and Obsolescence in America* (Cambridge, MA: Harvard University Press, 2006). For an interesting look at Apple's use of planned obsolescence in its design of the ever-popular iPod see www.cerge.cuni.cz/ pdf/events/papers/060410_t.pdf.

191   a molecular Band-Aid for progeria?

"Breakthrough in Premature Ageing," *New Scientist*, March 12, 2005; P. Scaffidi and T. Misteli. 2005. Reversal of the cellular phenotype in the premature aging disease Hutchinson-Gilford progeria syndrome. *Nat Med* 11(4):440–445.

194   big baby small pelvis

For more on childbirth from an evolutionary perspective see pages 183–203 in W. Trevathan, E. O. Smith, and J. J. McKenna, *Evolutionary Medicine* (New York: Oxford University Press, 1999); K. R. Rosenberg and W. R. Trevathan, "The Evolution of Human Birth," *Scientific American*, November 2001; H. Nelson, R. Jurmain, and L. Kilgore, *Essentials of Physical Anthropology* (St. Paul, MN: West Publishing, 1992).

196   *"Our ancestors entered the Pliocene"*

Elaine Morgan, personal communication. *The Descent of Woman* (New York: Stein and Day, 1972); E. Morgan, *The Aquatic Ape Hypothesis* (London: Souvenir Press, 1997); E. Morgan, *The Aquatic Ape: A Theory of Human Evolution* (London: Souvenir Press, 1982); E. Morgan, *The Scars of Evolution* (New York: Oxford University Press, 1994); E. Morgan, *The Descent of the Child: Human Evolution from a New Perspective* (New York: Oxford University Press, 1995); A. C. Hardy, "Was Man More Aquatic in the Past?" *New Scientist*, March 17, 1960; F. W. Jones, *Man's Place among the Mammals* (New York, London: Longmans, E. Arnold & Co., 1929); Kate Douglas, "Taking the Plunge," *New Scientist*, November 25, 2000. For an interview with Elaine Morgan see Kate Douglas, "Interview: The Natural Optimist," *New Scientist*, April 23, 2005.

201   *never really "got" what the theory was*

A. Kuliukas. 2002. Wading for food the driving force of the evolution of bipedalism? *Nutr Health* 16(4):267–289. See also Libby Brooks, "Come on in—the Water's Lovely," *Guardian*, May 1, 2003.

202   water babies

The study mentioned in the chapter is R. E. Gilbert and P. A. Tookey. 1999. Perinatal mortality and morbidity among babies delivered in water: surveillance study and postal survey. *BMJ* 319(7208):483–487. For a beautiful illustrated book with many photographs of mothers giving birth and swimming with their children see J. Johnson and M. Odent, *We Are All Water Babies* (Berkeley, CA: Celestial Arts Publishing, 1995); E. R. Cluett, R. M. Pickering, K. Getliffe, and N. J. St George Saunders. 2004. Randomised controlled trial of labouring in water compared with standard of augmentation for management of dystocia in first stage of labour. *BMJ* 328(7435):314; E. R. Cluett, V. C. Nikodem, R. E. McCandlish, and E. E. Burns. 2004. Immersion in water in pregnancy, labour and birth. *Cochrane Database Syst Rev* (2):CD000111. For the Italian study mentioned in the text see A. Thoeni, N. Zech, L. Moroder, and F. Ploner. 2005. Review of 1600 water births: does water birth increase the risk of neonatal infection? *J Matern Fetal Neonatal Med* 17(5):357–361. Water birthing is not without controversy; to see a study that found no positive correlation see K. Eckert, D. Turnbull, and A. MacLennan. 2001. Immersion in water in the first stage of labor: a randomized controlled trial. *Birth* 28(2):84–93.

204 episiotomy

As all medical procedures do, episiotomies show a marked variation in how often they are performed in different countries. For example, rates of episiotomies in the United States are still above 30 percent as compared to about 10 percent of births in northern Europe. For more details see S. B. Thacker and H. D. Banta. 1983. Benefits and risks of episiotomy: an interpretative review of the English language literature, 1860–1980. *Obstet Gynecol Surv* 38(6):322–338; for possible alternatives to episiotomies see M. M. Beckmann and A. J. Garrett. 2006. Antenatal perineal massage for reducing perineal trauma. *Birth* 33(2):159.

204 Dr. Myrtle McGraw and "water-friendly" behavior

M. B. McGraw, *The Neuromuscular Maturation of the Human Infant* (New York: Columbia University Press, 1943).

INDEX